T0331170

Transforming Gender-Based Healthcare with AI and Machine Learning

This book provides a thorough exploration of the intersection between gender-based healthcare disparities and the transformative potential of artificial intelligence (AI) and machine learning (ML) and covers a wide range of topics from fundamental concepts to practical applications.

Transforming Gender-Based Healthcare with AI and Machine Learning incorporates real-world case studies and success stories to illustrate how AI and ML are actively reshaping gender-based healthcare and offers examples that showcase tangible outcomes and the impact of technology in healthcare settings. The book delves into the ethical considerations surrounding the use of AI and ML in healthcare and addresses issues related to privacy, bias, and responsible technology implementation. Emphasis is placed on patient-centered care, and the book discusses how technology empowers individuals to actively participate in their healthcare decisions and promotes a more engaged and informed patient population.

Written to encourage interdisciplinary collaboration and highlight the importance of cooperation between health professionals, technologies, researchers, and policymakers, this book portrays how this collaborative approach is essential for achieving transformative goals and is not only for professionals but can also be used at the student level as well.

Studies in Intelligent Systems and Cognitive Computing

Series Editors: Vijender Kumar Solanki, Vinit Kumar Gunjan and Mohammed Usman

This new book series will contain books on theory, applications, and design methods of Intelligent Systems and Intelligent Computing. Virtually all disciplines such as engineering, natural sciences, computer and information science, ICT, economics, business, e-commerce, environment, healthcare, life sciences will be covered.

If you are interested in writing or editing a book for the series or would like more information, please contact Cindy Carelli, cindy.carelli@taylorandfrancis.com.

Transforming Gender-Based Healthcare with AI and Machine Learning
Edited by Meenu Gupta, Rakesh Kumar and Zhongyu Lu

For more information on this series, please visit: www.routledge.com/Studies-in-Intelligent-Systems-and-Cognitive-Computing/book-series/CRCSISCC

Transforming Gender-Based Healthcare with AI and Machine Learning

Edited by Meenu Gupta, Rakesh Kumar, and Zhongyu Lu

CRC Press
Taylor & Francis Group
Boca Raton London New York

CRC Press is an imprint of the
Taylor & Francis Group, an **informa** business

Designed cover image: Shutterstock—VectorMine

First edition published 2025
by CRC Press
2385 NW Executive Center Drive, Suite 320, Boca Raton FL 33431

and by CRC Press
4 Park Square, Milton Park, Abingdon, Oxon, OX14 4RN

CRC Press is an imprint of Taylor & Francis Group, LLC

ISBN: 978-1-032-75210-5 (hbk)
ISBN: 978-1-032-75322-5 (pbk)
ISBN: 978-1-003-47343-5 (ebk)

DOI: 10.1201/9781003473435

Typeset in Times
by Apex CoVantage, LLC

Contents

Somya Srivastava, Gaurav Dubey, Disha Mohini Pathak, Kamlesh Sharma, and Kuldeep Kumar

Chapter 12 Early Ethical Considerations and Societal Impacts on Gender Health 205

Antonitta Eileen Pious, K. Ananthi, Mary Fabiola, and Benedict Tephila

Preface

In a rapidly evolving healthcare landscape, integrating cutting-edge technologies such as artificial intelligence (AI) and machine learning (ML) can potentially revolutionize the way healthcare is delivered, especially in addressing gender-based health disparities. The proposed book explores the intersection of healthcare, technology, and gender equity. In this comprehensive guide, readers embark on a journey through the evolving healthcare landscape, where cutting-edge technologies like AI and ML play a pivotal role in revolutionizing healthcare delivery. The book begins by introducing the transformative potential of AI and ML in healthcare, setting the stage for a deep dive into their applications in gender-based healthcare. Readers gain insights into how AI is reshaping women's health diagnosis and treatment while discovering how ML enhances outcomes in men's health. Moreover, the book explores technology-driven approaches to create inclusive healthcare solutions that address the unique needs of diverse gender identities.

Data-driven insights take center stage as the book discusses predictive analytics for uncovering gender-based health trends and demonstrates how ML precision-tailors healthcare solutions based on gender-specific needs. Real-world success stories showcase the tangible impact of technology in gender-based healthcare, empowering patients and healthcare professionals alike. Furthermore, the book explores emerging trends, ethical considerations, and the future of gender-based healthcare in a digital age.

Transforming Gender-Based Healthcare with AI and Machine Learning is an indispensable resource for healthcare professionals, researchers, policymakers, and anyone passionate about harnessing technology to achieve equitable and effective healthcare for all genders. With its real-world examples, ethical discussions, and insights into emerging trends, this book illuminates the profound role of AI and ML in reshaping the healthcare landscape for the better.

About the Editors

Dr. Meenu Gupta is associate professor at the UIE-CSE Department, Chandigarh University, India. She is pursuing her Post Doc Fellowship from MIR Lab, USA. She completed her Ph.D. in Computer Science and Engineering from Ansal University Gurgaon, India, in 2020. She has more than 16 years of teaching experience. Her research areas cover machine learning, intelligent systems, and data mining, with a specific interest in artificial intelligence, image processing and analysis, smart cities, data analysis, and human/brain-machine interaction (BMI). She has edited more than 17 books and authored four engineering books. She reviews several journals, including *Big Data*, *Artificial Intelligence Review*, *CMC*, *Scientific Reports*, *Digital Health*. She is a life member of ISTE and IAENG. She is also a senior member of IEEE. She has authored or co-authored more than 37 book chapters and over 200 papers in refereed international journals and conferences. She also organised many conferences technically sponsored by the IEEE Delhi Section and AIP.

Dr. Rakesh Kumar is professor and associate director at the UIE-CSE Department, Chandigarh University, Punjab, India. He is pursuing his Post Doc Fellowship from MIR Lab, USA. He completed his Ph.D. in Computer Science and Engineering from Punjab Technical University, Jalandhar, 2017. He has more than 20 years of teaching experience. His research interests are IoT, machine learning, and natural language processing. He has edited more than seven books with the reputed publishers like Taylor & Francis Group and authored five books. He works as a reviewer for several journals, including *Big Data*, *CMC*, *Scientific Reports*, *TSP*, *Multimedia Tools and Applications*, and *IEEE Access*. He is a senior member of IEEE. He has authored or co-authored more than 170 publications in various national/international conferences and journals. He is also an organizer and editor of many international conferences under the aegis of IEEE and AIP.

Dr. Zhongyu Lu is a professor in the Department of Computer Science and is the research group leader of information and system engineering (ISE) at the Centre of High Intelligent Computing (CHIC). She was previously team leader in the IT department of the Charlesworth Group Publishing Company. She successfully led and completed two research projects in the area of XML database systems and document processing in collaboration with Beijing University. Both systems were deployed as part of company commercial productions. Professor Lu is UKCGE Recognised Research Supervisor (UK Council of Postgraduate Education). Professor Lu has published 11 academic books and more than 200 peer reviewed academic papers. Her research publications have 35,606 reads and 1008 citations by international colleagues, according to incomplete statistics from ResearchGate, Scopus, and Google Scholar. Professor Lu has acted as the founder of and a program chair for the International XML Technology Workshop for 11 years and serves as chair of various international conferences. She is the founder and editor-in-chief of *International Journal of Information Retrieval Research*, serves as a BCS examiner of Database and Advanced Database Management Systems, and is an FHEA. She has been the UOH principle investigator for four recent EU interdisciplinary (computer science and psychology) projects: Edumecca (student responses system) (143545-LLP-NO-KA3-KA3MP), DO-IT (multilingual student response system) used by more than 15 EU countries (2009–1-NO1-LEO05–01046), and DONE-IT (mobile exam system) (511485-LLP-1–2010-NO-KA3-KA3MP), HRLAW2016–3090 / 001–001.

Contributors

Shakeel Ahmed
College of Computer Sciences and
    Information Technology
King Faisal University
Al-Ahsa, Saudi Arabia

K. Ananthi
Sri Krishna College of Engineering and
    Technology
Coimbatore, India

Sharad Chauhan
Chitkara University Institute of
    Engineering and Technology
Chitkara University
Punjab, India

Gaurav Dubey
KIET Group of Institutions
Delhi, India

Priya Dubey
Amity School of Engineering &
    Technology
Amity University
Jharkhand, India

Mary Fabiola
Sri Krishna College of Engineering and
    Technology
Coimbatore, India

Fathimathul Rajeena P.P.
College of Computer Sciences and
    Information Technology
King Faisal University
Al-Ahsa, Saudi Arabia

Meenu Gupta
Department of Computer Science and
    Engineering
Chandigarh University
Punjab, India

Shubham Gupta
Model Institute of Engineering and
    Technology
Jammu, J&K, India

Md. Mehedi Hassan
Computer Science and Engineering
    Discipline
Khulna University
Khulna, Bangladesh

Anupriya Jain
School of Computer Applications,
    MRIIRS
Faridabad, India

Balaram Yadav Kasula
University of the Cumberlands
Williamsburg, Kentucky

Rohit Khankhoje
PT. Ravishankar Shukla University
Raipur, India
University of Texas
Austin, Texas

Harashleen Kour
Model Institute of Engineering and
    Technology
Jammu, J&K, India

Kuldeep Kumar
Amity University
Noida, Uttar Pradesh, India

Rakesh Kumar
Department of Computer Science and
    Engineering
Chandigarh University
Punjab, India

Vinod Kumar
University Institute of
    Engineering
Chandigarh University
Mohali, Punjab, India

Sonam Lata
School of Computer Science and
    Engineering
IILM University
Gurugram, Haryana, India

Manoj Kumar Mahto
Vignan Institute of
    Technology and Science,
    Deshmukhi (V),
    Pochampally (M)
Yadadri-Bhuvanagiri District
Telangana, India

Nandita Manchanda
Department of Computer Science and
    Engineering
Chandigarh University
Mohali, India

Syeda Husna Mehanoor
Department of
    Computer Science
College of Computer Sciences and
    Information Technology
King Faisal University
Al-Ahsa, Saudi Arabia

**Seyed Mahmoud Sajjadi
    Mohammadabadi**
University of Nevada, Reno
Computer Science and Engineering
    Department
Reno, Nevada

Mehrnaz Mostafavi
Shahid Beheshti University of Medical
    Sciences
Faculty of Allied Medicine
Tehran, Iran

Thamizhiniyan Natarajan
Department of Computer Science and
    Applications
The Gandhigram Rural Institute
    (Deemed to be University)
Gandhigram, India

Mrinal Paliwal
Department of Computer Science and
    Engineering
Chandigarh University
Mohali, India

Disha Mohini Pathak
Department of
    Computer Science
ABES Engineering College
Ghaziabad, India

Mahsa Borhani Peikani
University of Nevada, Reno
School of Medicine
Reno, Nevada

Shanmugavadivu Pichai
Department of Computer Science and
    Applications
The Gandhigram Rural Institute
    (Deemed to be University)
Gandhigram, India

Antonitta Eileen Pious
Sri Krishna College of Engineering and
    Technology
Coimbatore, India

Chander Prabha
Chitkara University Institute of
    Engineering and Technology
Chitkara University
Punjab, India

Faezehalsadat Seyedkhamoushi
Maharishi International University
Department of Management
Fairfield, Iowa

Kamlesh Sharma
Manav Rachna International
    Institute of Research and
    Studies
Faridabad, India

Palvi Sharma
Department of Computer
    Science and
    Engineering
Chandigarh University
Punjab, India

Seema Sharma
School of Computer Applications
    MRIIRS
Faridabad, India

Somya Srivastava
Department of Computer Science
ABES Engineering College
Ghaziabad, India

Benedict Tephila
Sri Krishna College of Engineering and
    Technology
Coimbatore, India

Pawan Whig
Vivekananda Institute of
    Professional Studies-TC
New Delhi, India

Nikhitha Yathiraju
University of the Cumberlands
Williamsburg, Kentucky

1 AI and Machine Learning in Modern Healthcare

Sharad Chauhan, Mrinal Paliwal,
Nandita Manchanda, and Meenu Gupta

1.1 INTRODUCTION

In recent decades, the healthcare sector has undergone a huge transformation with the integration of artificial intelligence (AI) and machine learning (ML) [1]. This will not only bring a new era of research and innovation but also significantly enhance the performance of delivery of healthcare services in an effective way. Far beyond being mere industry buzzwords, AI and ML have the capability to bring a revolutionary change in healthcare, starting from diagnosis of diseases, their effective treatment, and variety of medicines available to address patients effectively [2].

Due to recent growth in medical data and advancements in big data diagnosis techniques, healthcare data requires some innovative techniques to manage large amounts of data. AI and ML have brought a significant improvement in modern life and provide revolutionary change in medicine and the healthcare sector with the help of its latest technologies. These techniques help physicians reduce their workload and provide accuracy in tasks, predication of diseases and quality of care to patients. Machine learning techniques are used for identifying trends in medical data and are helpful in developing predication models which are useful in increasing efficiency of the healthcare system and managing electronic data in a better way. Recently we have seen the use of machine learning techniques in the COVID-19 pandemic where deep learning techniques are used for managing ventilators, tracking patients and their beds, rooms and even staff. AI techniques are also helpful in genetic sequencing and developing vaccines as well as monitoring patients [3].

ML techniques are helpful in extracting informative data from unstructured data and predicting results from the data which is very helpful in taking decisions for the medical supervisors. It is also useful in medical research where practitioners can get benefits for their medical trials. In today's scenario, remote monitoring of patients is also providing help to patients by getting online assistance from doctors. In the COVID-19 pandemic remote monitoring of patients was very helpful when in-person interaction between patients and doctors was not possible. In the current scenario, it is also useful for elderly people with mobility issues. Here remote monitoring is helpful in providing assistance to patients and improving the efficiency of the modern healthcare system [4].

AI- and ML-based techniques are helpful identifying diseases, drug discovery, identification of medical imaging, personalized remedies of medicines, behavioral modification, intelligently maintaining medical records, clinical trials, and research. It starts with data collection and management related to patients; this data identification its helpful in prediction of diseases; and based upon the medical history of patients it can recommend medicines. AI techniques can help new researchers in the medical field and discover new patterns and medicines to treat new kinds of diseases [5].

In this chapter, we will discuss the role of AI and ML techniques in the modern healthcare sector, as well as how these innovative techniques are helpful in improving the lives of medical practitioners and patients in the modern era. We will highlight the use of innovative techniques used for managing healthcare records and identifying diseases, providing on time medicines to patients and proving the effectiveness of the modern healthcare system.

1.2 RELATED WORK

T. Davenport and R. Kalakota (2019) have discussed the importance of artificial intelligence in the field of healthcare. They have shown how complex and large amounts of data are handled by companies by using latest technologies. These technologies are helpful in diagnosing diseases, their treatment and managing healthcare data. They have shown how AI-based techniques can reduce the workload of healthcare professionals by their innovative approaches. They have also highlighted how to manage patients, various administrative tasks, and research activities in the medical field [6].

S. Datta et al. (2020) have shown how AI is used for different healthcare and research purposes, starting from detecting of diseases to management of chronic conditions, and providing healthcare services. They also focused on the relationship of AI with the alginate which is part of tissue engineering field. Alginate is natural polymer which is a cell and used in tissue engineering [7].

J. Lorkowski et al. (2021) have shown the benefits of AI in medicines, health and the economy in the healthcare sector. In their review work, they have AI and ML models that are helpful in medical science. They have also talked about blockchain technologies, medical science sensitive data, and medical documentation. These innovative technologies are helpful in patient's wellbeing [8].

A. Bohr and K. Memarzadeh (2021) have shown that due to increasing demand of healthcare services many countries are having a shortage of medical physicians. They have shown that due to the development of the latest technologies, patient's expectations and level of service requirements are also increasing. So, fulfilling the expectations of patients in the modern era requires the latest innovative technologies. AI-based tools provide cutting-edge healthcare offerings that are improving and reducing the cost of healthcare services [9].

M. Javaid et al. (2022) have shown the benefits of machine learning in the healthcare sector. They have highlighted how machine learning improves the accuracy and speed of the work doctors do. As demands on physicians increase in all countries, ML techniques can minimize their workload by managing healthcare data and minimizing errors. These techniques can be helpful in early detection of pandemics, which can lead to the curing of diseases in a timely manner. ML-based tools are minimizing cost in the healthcare sector, providing treatment of different kinds of diseases and improving efficiency in the healthcare sector [10].

T.M. Ghazal, et al. (2021) have highlighted the value of IoT-based sensor networks for healthcare applications. They have highlighted how machine learning approaches are helpful in implemented IoT-based sensor networks, which are useful in managing large amounts of healthcare data efficiently. In their work they have proposed how AI-based techniques are applied in WSNs which are beneficial for healthcare peoples [11].

S. Siddique and J.C.L. Chow (2021) have shown that ML and AI techniques have been used for those tasks that require human intelligence. Healthcare communication is a very important aspect for providing important facts to patients and informing them about healthcare services. In their proposed work they have proven how ML is proven to be effective in managing complex problems and providing benefits to patients. They cite various examples related to the healthcare sector, such as cancer therapy, COVID-19 health education etc. [12].

M.V. Schaar et al. (2020) have shown how ML and AI have proven beneficial in the COVID-19 pandemic. This disease has burdened the healthcare system and existing infrastructure and resources are not feasible for dealing with such kind of diseases. Integration of innovative techniques of AI and ML helped in the COVID-19 pandemic and saved may lives [13].

M.Y. Shaheen (2021) has shown how AI provides support in modern healthcare systems by providing surgery assisting robots who can grasp the medical concepts better than human beings and detect minor patterns that are manually not feasible. They have focused mainly on clinical trials, medical drug discovery and support to patients. This AI technology based clinical research work has been helpful in pharma companies for improving drug discovery and analyzing their targets. These innovative methods are helpful in providing medical assistance to patients and their medical data at every level, which can help in growing the healthcare sector at large scale [14].

A.B. Montero et al. (2021) has highlighted the use of AI in medical image analysis. In the modern healthcare sector, image analysis techniques are very helpful in radiology and pathology where x-ray images are used. It is helpful in areas like diagnosis of diseases and classification of patterns in the medical field. They have also highlighted new trends and research directions in medical science with the use of AI [15].

1.3 RELATIONSHIP AMONG AI, NLP, ML AND DL

AI came in existence in 1951. When the first AI program was developed its prime objective was for research activities. In the beginning, AI research focused on rule-based and expert systems, and in later stages it moved to making machines smarter. AI is a broad and evolving branch of computer science whose task is to create smarter machines that can perform tasks in intelligent ways. In today's scenario, AI is not limited only to research but has impacted other areas like transportation, digital finance and the healthcare sector. In the recent past a continuous development has been observed in the clinical sector where AI-based models are used for patient care.

AI techniques include ML, deep learning (DL) and natural language processing (NLP), and each of these technologies have a well-defined functionality. Figure 1.1 depicts the relationship among these technologies. NLP is subfield of AI that deals

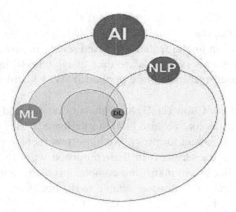

FIGURE 1.1 Relationship among AI, ML, DL and NLP.

with interpreting and generating human languages. NLP provides helps in classification, categorization and sentimental analysis [16].

ML algorithms, which is also subfield of AI, are based on helping machines learn from data and improving their efficiency leading to improved performance of the overall system. Machines learn from experiences by using mathematical techniques that help machines to further improve. ML is already used in the healthcare sector and requires more future considerations in this sector [17].

DL is a subfield of ML that uses models based on deep neural networks need to identify patterns with little human effort. Deep learning is applied to large amounts of data and extracts useful patterns and information, which can take human decades if done manually. DL methods use hierarchical levels of artificial neural networks that are motivated by the biological human brain. DL is used to build new functionality that not only discovers new features from a big data set but also forms new ideas [18]. In the healthcare sector it is not only used to detect diseases but also predict models for treating these diseases. DL techniques are also beneficial in cancer detection, radiology and pathology where they can provide treatment by building models that cannot be done easily by human beings [19].

The way AI and ML technologies are integrated marks a revolutionary shift in the healthcare industry through innovative methods and new advancements. This paradigm shift in technology is not only based on theoretical concepts but also in practical implementation in medical research, diagnostics and improved therapies. Improvements in the latest innovative techniques based on these technologies in the medical sector will replace the conventional methods of curing diseases, therapeutics and patient care.

1.4 DIAGNOSTIC PRECISION AND TREATMENT PLANNING

1.4.1 EARLY DISEASE DETECTION

Research has shown that AI and ML algorithms can be used for better analyzing large datasets, identifying similar patterns and generating insights aid in precise

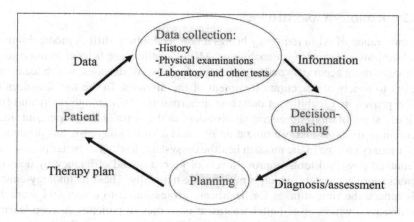

FIGURE 1.2 Early disease detection and treatment process.

and early detection of diseases [20]. These technologies are capable enough for interpreting medical imaging such as MRIs and CT scans and their complex data, which leads to predicting the diseases at the early stages and helps medical practitioners accurately diagnose diseases. Research has also indicated that AI technologies play a significant role in early disease detection in various diseases like cancer, diabetes and cardiovascular disorders. This analytical capability of these technologies will be exceptionally helpful in early disease detection and minimize the risk of life to patients, improves outcomes [21]. These innovative technologies provide a powerful toolset to medical professionals that helps them cure diseases at early stages, prescribe medicines to patients and overall improve patient care, leading to the overall improvement in the healthcare system. Some advancements of AI in the medical field are recurrent neural networks (RNNs), which help identify dependencies in health-related data. Other predictive models using AI techniques in medical field include the random forest and ensemble methods [22].

Figure 1.2 shows the process of disease detection and treatment. The first step is to collect data from the patients, in which the medical history of a patient is considered and based on that all examination and testing is performed. These test results and medical history will be helpful in diagnosis of the diseases and their treatment. After that doctors will make decisions related to the treatment plans and therapy plans that will be helpful for patients in curing the diseases. The whole process is very carefully examined in Figure 1.2.

Machine learning algorithms are very helpful in certain kinds of skin diseases. Most skin cancer deaths are caused by a type of skin cancer known as melanoma. This kind of disease is prevented by early detection and timely treatment. Management includes carefully checking images with the use of machines, which includes high resolution cameras. Proper use of these high-resolution images is very helpful in identifying skin diseases and after that machine learning applies to extract proper information from these images, which are helpful in managing this disease. Further image segmentation applies to extract skin related issues. Certain kinds of parameters are also taken to find patterns that are helpful in treatment of melanoma [23].

1.4.2 RADIOLOGY AND IMAGING

The integration of AI in radiology brings a new technology shift in medical images, which improves diagnostics capabilities. ML algorithms are helpful in interpreting the images more accurately and avoiding misdiagnosis of diseases, which leads radiologists to timely and accurate treatment of the diseases. In the last few decades, AI has proven its capability of detecting abnormalities in mammography and finding early signs of neurodegenerative disorders in the brain. This proper and timely detection of diseases and minimization of the risk of misdiagnosis has produced a revolutionary change in the modern healthcare system, leading to better patient care. AI-enabled powerful tools improve levels of precision and efficiency by detecting diseases more accurately, which is not possible manually. These technology-enabled tools reduce the time it takes for healthcare professionals to detect and accurately cure diseases, which gives better treatment outcomes to patients. A deep learning model named convolutional neural networks (CNNs) is used in finding patterns and anomalies in radiological images [24]. This will bring a technological improvement in redefining the diagnostic accuracy in medical imaging.

1.5 PERSONALIZED MEDICINES AND TREATMENT PLANS

1.5.1 PERSONALIZED MEDICINE: TAILORING TREATMENT FOR INDIVIDUALS

AI-based approaches in the field of pharmaceuticals have shown a revolutionary change by providing genetic medicines to individuals that provide proper treatment plans in the healthcare sector. These AI-based approaches predict how patients will responds to specific medicines. These will focus on analysis of genetic variations, and ML model pattern recognition will predict how patients will respond, minimizing adverse effects, improving therapy efficiency, which represents a significant improvement in personalized medicines.

This capability not only minimizes the risk of undesirable results but also improves the therapy outcomes by proper alignment of the treatment plans with the patient's report generated by AI approaches. This will provide a paradigm shift in the pharmaceutical field by enabling healthcare professionals to provide customized treatment plans to patients based on their diseases and offering them a modern healthcare infrastructure. This personalized medicine approach in will represent a significant change in precision medicine, which provides an effective and secure way of supplying medicine that are helpful in the treatment of individuals and act as a new and modern way of patient-oriented healthcare. Reinforcement learning algorithms are very helpful in personalized therapies. These algorithms are dynamically changing the treatment plans for patients based upon their recovery and responses to the medicines, which is helpful for medical practitioners to take medical decisions related to patients. This will be helpful in providing customized treatment plans for patients and take real-time interventions in patient care.

1.5.2 TREATMENT RECOMMENDATIONS AND PLANS

In the last few decades, it has been identified that ML-based algorithms are very effective in modern healthcare systems. They can analyze medical data based on a patient's diseases, which is helpful in clinical trials and aids practitioners in medical research, which leads to effective treatments. They can analyze patterns that guide healthcare professionals to properly monitor health profiles of patients and give them proper treatment plans. This boosts doctors' decision-making capabilities and deliver care according to a patient-centric healthcare model. This decision-making ability of doctors will produce better treatment recommendation significantly improve patients' health, which optimizes medical outcomes [25]. This will also minimize the side effects of medicines given to patients, as these will be based on AI technology. So, integration of ML in healthcare decision-making will be a revolutionary step for providing personalized medicines in an effective and precise way.

1.5.3 DRUG DISCOVERY ACCELERATION

AI-based algorithms are very important in drug discovery where these technologies can identify potential drug candidates. In this area the mixture of quantum computing and machine learning algorithms are used to identify the molecular interactions and predict their properties, which is not easy with classical approaches. This will be helpful in drug development processes based on the identified molecular properties. This will bring a revolutionary change in research and development in the pharmaceutical field.

1.6 PATIENT CARE AND OPERATIONAL EFFICIENCY

1.6.1 PREDICTIVE ANALYSIS FOR PATIENT CARE

The effective use of AI and ML will result in better treatment plans for patients and improve patient care. This will enhance the operational efficiency of hospitals by providing a patient-oriented healthcare system that includes AI-enabled tools for monitoring a patient's admission in hospitals, allocating healthcare resources effectively among patients, providing suitable medicines, and properly managing data. This will reduce waiting times for patients and provide timely delivery of healthcare facilities to patients.

These AI-based tools efficiently analyze patients' historical data, which helps obtain valuable insights that can be used to treat diseases. It is important to understand that these technology-based tools will not replace healthcare professionals; this will be for their help in managing patients. This will enable hospitals to allocate staff and resources among patients efficiently, preventing shortages of medical resources and ensure timely and efficiently delivery of care among patients, which leads to improved patient care. Healthcare professionals also get benefits by properly focusing on the patients and giving them proper treatment as all information they want related to patients will be delivered to them in a timely manner with the help of AI tools. These technology-enabled models are capable of predictive analysis and will help modern hospitals allocate medical resources properly, reduce patient waiting time, generate a more responsive system, and pay more attention to patients, which leads to a patient-centered modern healthcare environment.

1.6.2 MODERN SUPPLY CHAIN MANAGEMENT

For proper delivery of healthcare services, modern healthcare systems require a proper supply chain management system. AI is proven to be a crucial component in this by its predictive analytic capability which properly forecasts inventory requirements and utilization of medical instruments in the healthcare sector. It properly examines the history of consumption medical accessories, detects patterns, continuously checks demand fluctuation and other data. Based upon this, it can make future requirements and ensure timely delivery of medical procurements. It not only checks stock requirements like shortages of procurement or overflow of stocks but also monitors the allocation of resources among different healthcare departments, which leads to timely delivery of medical supplies among all departments. This streamlined process will minimize the risk of shortages of medical accessories and enhance the patient care and operational efficiency of the modern healthcare system. The role of AI in supply chain management will bring a new era of technology shift which fulfils all dynamic resource requirements of the modern healthcare industry.

1.7 ETHICAL CONSIDERATIONS AND CHALLENGES IN THE HEALTHCARE INDUSTRY

As AI and ML technologies brings a significant change in the healthcare sector but it is important to understand that these innovative methods will be gaining progress without facing any challenges. These challenges include security and data privacy issues with some ethical considerations. As technology grows with innovations there always be a responsibility to ensure that it must follow all standards and always preserve privacy. Following these standards will improve the healthcare system and ensure that it is not compromising security and privacy. We cannot trade security and privacy in exchange for the benefits of technologies and advancements. Getting better patient care, improving diagnostic precision and performance outcomes requires a valid approach for ethical concerns. As technology grows, ethical considerations are required when considering data privacy, algorithmic bias, and proper refinement of new advancements. The use of AI and ML technologies in healthcare ensures that these technologies follow the regulatory and ethical standards whose prime objective is to foster patient wellbeing.

By following these considerations, we can get the proper benefits of technologies in the healthcare sector with full accountability. These ethical principles will serve as guidelines for healthcare systems that are not only technology enabled but also provide a trustworthy and compassionate system for all medical professionals. These principles and technologies will open an innovative path for a future in which these new techniques and algorithms will benefit individuals and the medical community by creating a patient-oriented and modern healthcare ecosystem.

1.7.1 SECURITY AND DATA PRIVACY

In the past few decades, digitization has played an important role in modern healthcare systems, ensuring security and privacy when it comes to patient data. An electronic health record of a patient will be enabled by innovative technologies of AI and ML, which makes a digital healthcare system efficient for maintaining sensitive medical data. Electronic health records will ensure prevention of sensitive data from unauthorized persons and maintain the secrecy of data. It is always important that patient data not be shared with others without their consent. Healthcare data must be enabled with innovative techniques that always align with data privacy and security standards.

In the healthcare sector, it is the responsibility of hospitals to maintain the trust of their patients that their health data follows measures of confidentiality and integrity. Ethical standards properly define data usage boundaries and proper access to sensitive data, preventing unauthorized access to data. AI techniques provide transparency in healthcare systems, which engenders a belief among patients and healthcare providers that their data is secure. So properly addressing all ethical standards containing privacy will ensure that technology advancements do not compromise standards and maintain patient trust and confidentiality in the modern healthcare industry.

1.7.2 Bias in Artificial Intelligence Algorithms

In the recent era, it has been shown that biases in AI algorithms generate different kind of challenges in the healthcare sector, which suffers from the limited decision-making capability of doctors. These biases include use of diversifying datasets, using fairness-aware algorithms and not properly following the validation process. Due to biased decisions related to prediction of diseases, treatment often suffers. These should be properly managed for addressing healthcare disparities by properly training the algorithms before implementation in the healthcare sector. This also affects the treatment recommendation given by doctors and diagnostic accuracy of the system. So as the healthcare sector is directly related to the life of patients, so these issues and challenges should be properly addressed. A proper unbiased mixing of AI and ML technologies should be required in the healthcare industry for advancing and ensuring expected outcomes without existing challenges. It is very important to understand how AI algorithms make decisions in critical medical situations.

1.8 IMPACT OF AI IN MODERN HEALTHCARE

The main aspects of AI and ML in the modern healthcare sector are shown in Figure 1.3. The main functionalities of these technologies in the healthcare sector are given as [26]:

- **Personalized Medicines:** AI techniques are efficient enough to analyze individual data and based upon that find out the diseases risk, proper treatment plans, and optimized use of medicines. Also, it suggests proper preventive actions against the diseases.
- **Advanced Diagnostic Techniques:** AI-based efficient tools are capable enough to diagnose diseases in early stages, which is very helpful in certain kinds of critical diseases like cancer and chronic diseases. They provide scanning facilities to patients before getting treated by the doctor.

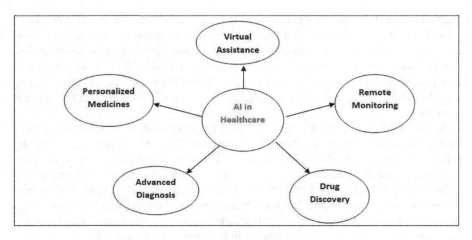

FIGURE 1.3   Impact of AI in healthcare.

- **Virtual Assistance:** AI enabled chatbots will provide virtual assistance to patients and offer solutions to their queries by simple answers, making assessments, minimizing the workload of doctors, and properly managing patients. This will work like providing basic mental support to patients.
- **Remote Monitoring:** AI-based wearable devices and sensor-enabled devices can remotely monitor the health of patients in the absence of medical supervisors. These devices are capable enough to monitor basic health conditions of patients such as temperature, hearth beats, calories count, blood sugar level etc. This will provide early signs of diseases to patients and based on this, patients can take proper treatment at early stages.
- **Drug Development and Discovery:** AI algorithms are helpful in uncovering genomics and drug data, which helps medical practitioners in their medical trials to identify new drugs, thereby improving the drug discovery process.

1.9 FUTURE PROSPECTS: A GLIMPSE INTO TOMORROW'S HEALTHCARE

Proper refinement of AI and ML algorithms, advancements in security and privacy, and a combined effort by the healthcare sector and technological developers will provide a revolutionary change in tomorrow's healthcare system and achieve greater milestones. This revolutionary change will result in a promising healthcare system with the integration of AI and ML technologies that provide more effective, personalized, and expert medical solutions, which will benefit individuals as well as society at large level. These changes will be more effective with the integration of emerging technologies like edge computing, federated learning, and explainable AI (XAI).

Explainable AI (XAI) provides a set of tools and a framework that helps in addressing the challenges in complex algorithms. This will be helpful in interpreting the prediction of ML models. This will provide transparency and maintain trust, which is important in the healthcare domain.

Federated learning will be useful for addressing security and privacy issues by building a common machine learning models without sharing data. It acts as a collaborative model that enables ML models to train from distributed datasets. It is helpful in the healthcare sector where the sensitive data of patients are distributed across different locations.

Edge computing is also helpful in the healthcare domain as the requirement of real-time data has increased steadily over the last few decades. It can enable a machine to access and analyze real-time data at source, which leads to increases in efficiency and minimizes the latency of AI applications in the medical domain.

These future technologies will enhance the capabilities of AI and ML in the healthcare sector by minimizing the limitations of the current scenario. Integration of these tools will result in a more powerful healthcare system. These technologies act as future frontiers that enhance innovation and provide better service in the healthcare sector.

1.10 CONCLUSION

The growth of healthcare sector has been increasing rapidly by use of AI and ML techniques. These technologies streamline the process of a patient's diagnosis, identifying patterns among large datasets, which aids in managing patient data and treatment, improvement in drug discovery processes and the overall efficiency of the medical system. They will provide better and simpler treatment plans for patients. Merging AI techniques with human effort will provide a new milestone in shaping the future of the healthcare sector. These technologies will provide a proper balance of technological advancement in the medical sector, which can lead to better medical solutions for both society and individuals. They are the future of the modern healthcare system whose aim is to provide more effective, personalized, and accessible solutions for patient care. These technologies will not compromise ethical standards and the privacy of patient's data. In the future more advancements in these technologies will continue to impact the modern healthcare system and better patient care.

REFERENCES

[1] R. Kaur, R. Kumar, and M. Gupta, "Review on Transfer Learning for Convolutional Neural Network," in *2021 3rd International Conference on Advances in Computing, Communication Control and Networking (ICAC3N)*, 2021, pp. 922–926.
[2] S. Chauhan, K. Pahwa, and S. Ahmed, "Telemedical and Remote Healthcare Monitoring Using IoT and Machine Learning," in *Computational Intelligence in Healthcare*, Boca Raton: CRC Press, 2022, pp. 47–66.
[3] H. Habehh and S. Gohel, "Machine Learning in Healthcare," *Curr. Genomics*, vol. 22, no. 4, pp. 291–300, 2021, doi: 10.2174/1389202922666210705124359.
[4] S. A. S. Chauhan and K. Pahwa, "Telemedical and Remote Healthcare Monitoring Using IoT and Machine Learning," in *Computational Intelligence in Healthcare Application, Challenges, and Management*, Boca Raton, London, New York: CRC Press, Taylor and Francis Group, 2023, p. 20.

[5] R. A. S. Chauhan and N. Arora, "IoT and Machine Learning Based Smart Healthcare System for Monitoring Patient," in *Artificial Intelligence of Health-Enabled Spaces*, 1st Edition, Boca Raton, London, New York: CRC Press, Taylor and Francis Group, 2023, pp. 1–19.

[6] T. Davenport and R. Kalakota, "The Potential for Artificial Intelligence in Healthcare," *Futur. Healthc. J.*, vol. 6, no. 2, pp. 94–98, Jun. 2019, doi: 10.7861/futurehosp.6-2-94.

[7] S. Datta, R. Barua, and J. Das, "Application of Artificial Intelligence in Modern Healthcare System," in *Alginates—Recent Uses of This Natural Polymer*, vol. 6, London: Ml, IntechOpen, 2020, pp. 130–137.

[8] J. Lorkowski, O. Grzegorowska, and M. Pokorski, "Artificial Intelligence in the Healthcare System: An Overview," in M. Pokorski (eds.), *Best Practice in Health Care. Advances in Experimental Medicine and Biology*, vol. 1335, Cham: Springer, 2021, pp. 1–10, doi: 10.1007/5584_2021_620

[9] A. Bohr and K. Memarzadeh, *The Rise of Artificial Intelligence in Healthcare Applications*. INC, 2020.

[10] M. Javaid, A. Haleem, R. Pratap Singh, R. Suman, and S. Rab, "Significance of Machine Learning in Healthcare: Features, Pillars and Applications," *Int. J. Intell. Networks*, vol. 3, no. February, pp. 58–73, 2022, doi: 10.1016/j.ijin.2022.05.002.

[11] T. M. Ghazal *et al.*, "IoT for Smart Cities: Machine Learning Approaches in Smart Healthcare—A Review," *Futur. Internet*, vol. 13, no. 8, 2021, doi: 10.3390/fi13080218.

[12] S. Muthukumarasamy, A. K. Tamilarasan, J. Ayeelyan, and M. Adimoolam, "Machine Learning in Healthcare Diagnosis," in *Blockchain and Machine Learning e-Healthcare Systems*, pp. 343–366, 2021, doi: 10.1049/pbhe029e_ch14.

[13] M. van der Schaar *et al.*, "How Artificial Intelligence and Machine Learning Can Help Healthcare Systems Respond to COVID-19," *Mach. Learn.*, vol. 110, no. 1, pp. 1–14, 2021, doi: 10.1007/s10994-020-05928-x.

[14] M. Y. Shaheen, "Applications of Artificial Intelligence (AI) in Healthcare: A Review," no. September, pp. 0–2, 2021, doi: 10.14293/S2199-1006.1.SOR-.PPVRY8K.vl.

[15] A. Barragán-Montero *et al.*, "Artificial Intelligence and Machine Learning for Medical Imaging: A Technology Review," *Phys. Medica*, vol. 83, no. May, pp. 242–256, Mar. 2021, doi: 10.1016/j.ejmp.2021.04.016.

[16] S. A. Alowais *et al.*, "Revolutionizing Healthcare: The Role of Artificial Intelligence in Clinical Practice," *BMC Med. Educ.*, vol. 23, no. 1, pp. 1–15, 2023, doi: 10.1186/s12909-023-04698-z.

[17] S. Chauhan, R. Arora, and N. Arora, "Researcher Issues and Future Directions in Healthcare Using IoT and Machine Learning," in *Smart Healthcare Monitoring Using IoT with 5G*, Boca Raton: CRC Press, 2021, pp. 177–196.

[18] D. Chatha, A. Aggarwal, and R. Kumar, "Comparative Analysis of Proposed Artificial Neural Network (ANN) Algorithm with Other Techniques," in *Research Anthology on Artificial Neural Network Applications*, IGI Global, 2022, pp. 1218–1223.

[19] D. Mitra, A. Paul, and S. Chatterjee, "Machine Learning in Healthcare," *AI Innovation in Medical Imaging Diagnostics*, pp. 37–60, 2021, doi: 10.4018/978-1-7998-3092-4.ch002.

[20] M. Gupta, R. Kumar, and A. Abraham, "Adversarial Network-Based Classification for Alzheimer's Disease Using Multimodal Brain Images: A Critical Analysis," *IEEE Access*, vol. 12, pp. 48366–48378, 2024, doi: 10.1109/ACCESS.2024.3381956.

[21] K. Varshney and M. Paliwal, "Heart Disease Diagnosis by Neural Networks," *J. Pharm. Res. Int.*, pp. 202–208, Oct. 2021, doi: 10.9734/jpri/2021/v33i46A32858.

[22] M. Ferdous, J. Debnath, and N. R. Chakraborty, "Machine Learning Algorithms in Healthcare: A Literature Survey," in *2020 11th Int. Conf. Comput. Commun. Netw. Technol. ICCCNT 2020*, 2020, doi: 10.1109/ICCCNT49239.2020.9225642.

[23] S. Gupta, N. Sharma, R. Tyagi, P. Singh, A. Aggarwal, and S. Chawla, "Cognitive Inspired and Computationally-Intelligent Early Melanoma Detection Using Feature Analysis Techniques," *J. Artif. Intell. Technol.*, Oct. 2023, doi: 10.37965/jait.2023.0334.

[24] M. Malik *et al.*, "Waste Classification for Sustainable Development Using Image Recognition with Deep Learning Neural Network Models," *Sustainability*, vol. 14, no. 12, p. 7222, Jun. 2022, doi: 10.3390/su14127222.

[25] M. Saxena and S. Ahuja, "Comparative Survey of Machine Learning Techniques for Prediction of Parkinson's Disease," in *2020 Indo—Taiwan 2nd International Conference on Computing, Analytics and Networks (Indo-Taiwan ICAN)*, Feb. 2020, pp. 248–253, doi: 10.1109/Indo-TaiwanICAN48429.2020.9181368.

[26] M. Gupta, S. Ahmed, R. Kumar, and C. Altrjman, *Computational Intelligence in Healthcare: Applications, Challenges, and Management.* Milton Park, Abingdon, Oxon: CRC Press, 2023.

2 Revolutionizing Gender-Specific Healthcare
Harnessing Deep Learning for Transformative Solutions

Pawan Whig, Balaram Yadav Kasula, Nikhitha Yathiraju, Anupriya Jain, and Seema Sharma

2.1 INTRODUCTION

Orientation explicit medical services remain as a basic space inside the more extensive range of clinical practice. Perceiving and tending to the extraordinary wellbeing needs and aberrations between sexual orientations is key to accomplishing even-handed and compelling medical services results. Lately, the union of orientation explicit medical services and state of the art innovation, especially deep learning, has proclaimed another time of extraordinary arrangements ready to upset the scene of medical services [1]. This chapter explores the convergence of gender-specific healthcare and deep learning technologies, examining the various ways in which advanced algorithms and data-driven approaches are transforming the delivery of healthcare tailored to the unique needs and characteristics of different genders [2].

Customary medical care practices have frequently moved toward therapy and diagnostics according to a summed-up point of view, ignoring the inborn contrasts in the science, physiology, and wellbeing encounters between sexual orientations. Notwithstanding, with the appearance of deep learning, a subset of man-made consciousness (simulated intelligence) that gains and iteratively improves from information, there exists an exceptional chance to unwind complex examples and subtleties well defined for orientation-based wellbeing [3]. Deep learning calculations, energized by huge datasets and computational power, can investigate multifaceted natural, hereditary, and social markers. This empowers the identification of orientation explicit wellbeing patterns, forecast of sickness dangers, and improvement of customized intercessions that consider the novel physiological and socio-social angles related with various genders [4].

All through this chapter, we will dig into different utilizations of deep learning in orientation explicit medical services, enveloping regions, for example, we will ask how deep learning calculations dissect standards of conduct and emotional wellbeing information to make customized mediations, representing orientation explicit psychological wellness variations.

  DOI: 10.1201/9781003473435-2

TABLE 2.1
Literature Review with Research Gap

Reference	Summary	Research Gap
[1]	Empowering women in medical activities from Prophet Muhammad's era to modern times in Pakistan.	Limited exploration of specific challenges faced by women in modern medical activities and potential solutions.
[2]	Deep learning applications in pediatric neuroimaging.	Lack of discussion on the challenges and limitations of deep learning in pediatric neuroimaging.
[3]	Public leadership for gender equality with a framework for policy change.	Limited insights into the practical implementation challenges and outcomes of gender transformative policy change.
[4]	Student dropout analysis and retention in higher education using AI and machine learning.	Need for more exploration on the specific factors leading to student dropout and the effectiveness of AI-based retention strategies.
[5]	Digital inclusion in agri-food systems in Bangladesh.	Limited exploration of challenges faced by farmers in adopting digital technologies and the potential impact on agriculture.
[6]	Proceedings of the Women in Medicine Summit: An evolution of empowerment.	Lack of detailed analysis on the evolution of empowerment in women in medicine and the specific areas of improvement needed.
[7]	Deep learning applications in bone mineral density estimation, spine vertebra detection, and liver tumor segmentation.	Limited discussion on the practical challenges and potential biases associated with deep learning in medical image analysis.
[8]	AI in precision medicine.	Need for more case studies and practical insights into the integration of AI in precision medicine for better patient outcomes.
[9]	Islamic finance solutions for environmental, social, & economic challenges in the digital age.	Limited exploration of the specific applications and challenges of Islamic finance solutions in addressing digital-age challenges.
[10]	Women in mechanical circulatory support: She persisted!	Need for more insights into the challenges faced by women in the field of mechanical circulatory support and their persistence.
[11]	Impact of industrialization on social mobility in rural communities.	Lack of exploration into specific measures and strategies for achieving inclusive and sustainable economic transformation.
[13]	Transformational IoT sensing for air pollution and thermal exposures.	Limited discussion on the societal impact and potential challenges of implementing IoT sensing for environmental monitoring.
[14]	Innovating for the circular economy: Driving sustainable transformation.	Lack of detailed exploration of specific innovations driving circular economy practices and their potential challenges.
[15]	Transforming the Toilet 6. Bathroom Battlegrounds: How public restrooms shape the gender order.	Need for more insights into the societal and cultural factors influencing public restrooms and shaping gender orders.

(Continued)

TABLE 2.1 (*Continued*)
Literature Review with Research Gap

Reference	Summary	Research Gap
[16]	MSME Access to Finance: The Role of Digital Payments.	Limited exploration of the challenges faced by micro, small, and medium enterprises (MSMEs) in accessing finance through digital payments.
[17]	Innovative knowledge generation: Exploring trends in the use of early childhood education apps in Chinese families.	Need for more insights into the effectiveness of early childhood education apps and potential challenges faced by Chinese families.

This chapter aims to elucidate the potential of deep learning as a catalyst for advancing gender-specific healthcare, while also addressing the ethical considerations, challenges, and future implications associated with integrating these technologies in the pursuit of equitable and effective healthcare for all genders.

Table 2.1 summarizes the main findings and research gaps across the diverse literature, providing a comprehensive overview of the identified gaps in each reference.

2.2 EVOLUTION OF GENDER-SPECIFIC HEALTHCARE

The evolution of gender-specific healthcare represents a progressive shift in medical practice, acknowledging and addressing the distinctive health needs and disparities that exist between different genders [5]. Throughout history, healthcare approaches were often generalized, overlooking the biological, physiological, and socio-cultural differences inherent in diverse genders. However, the evolution of gender-specific healthcare has traversed several pivotal stages, shaping the way healthcare is delivered and tailored to meet the unique needs of individuals across genders.

1. Historical Recognition of Gender Differences: Early medical practices acknowledged some gender-specific health disparities, such as disparities in reproductive health or prevalent diseases.
2. Integration of Orientation explicit Rules: Administrative bodies and medical care associations began coordinating orientation explicit rules into clinical practice. These guidelines aim to address healthcare disparities by focusing on gender-based differences in diagnostics, treatments, and preventive care [6,7].
3. Advancements in Precision Medicine: The emergence of precision medicine has ushered in a new era, emphasizing personalized healthcare tailored to an individual's unique genetic makeup, lifestyle, and environmental factors. Orientation explicit considerations turned into a necessary piece of customized treatment plans [8].
4. Acknowledgment of Socio-social Elements: Contemporary ways to deal with orientation explicit medical services presently underline the interconnection of wellbeing, taking into account organic contrasts as well as

socio-social and conduct factors affecting wellbeing results across sexes. This acknowledgment has incited a more all-encompassing way to deal with medical care [9,10].

5. Continued Spotlight on Wellbeing Value: Currently, there is an increased focus on achieving health equity, with a goal to eliminate gender-based health disparities. Efforts are directed toward promoting equal access to healthcare, reducing biases in medical research, and fostering inclusive and culturally competent care.

2.3 ROLE OF DEEP LEARNING IN HEALTHCARE TRANSFORMATION

The job of deep learning in medical services change is significant, upsetting different parts of the clinical field by utilizing progressed calculations and information driven systems [11]. Deep learning, a subset of man-made reasoning (man-made intelligence) zeroed in on emulating the human mind's brain organizations, has arisen as an impetus for groundbreaking changes, especially in medical services as show in Figure 2.1.

Deep learning models succeed in breaking down immense datasets, pictures, and complex clinical records to aid diagnostics. These models display excellent accuracy in deciphering clinical imaging, for example, X-ray checks, X-beams, and pathology slides, frequently equaling or outperforming human skill. Deep learning speeds up drug discovery by examining atomic designs, identifying potential medication applicants, and anticipating their adequacy. This innovation smooths out the course of medication improvement, lessening time and expenses related to offering new prescriptions to the public. Deep learning adds to customized medication by fitting treatment plans in light of individual patient information. These models think about hereditary varieties, biomarkers, and patient-explicit variables to suggest altered treatments, streamlining treatment results. Utilizations of deep learning in medical care activities improve effectiveness via computerizing authoritative undertakings, smoothing out work processes, and advancing asset designation. Predictive analytics aids in comprehensive hospital management by reducing wait times and improving patient flow. Deep learning calculations work with the examination of unstructured information in electronic wellbeing records, separating significant experiences from

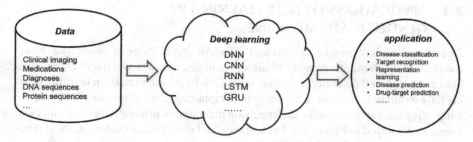

FIGURE 2.1 Deep learning in healthcare transformation.

clinical notes, reports, and patient accounts. This facilitates navigation, exploration, and understanding of patient instructions. Deep learning enhances remote patient monitoring by analyzing data from wearable devices and sensors. These models can detect anomalies, monitor vital signs, and provide real-time insights for remote healthcare management.

2.3.1 UNDERSTANDING DEEP LEARNING

Deep learning is a subset of man-made brainpower (simulated intelligence) that includes preparing calculations known as brain organizations to perceive examples and make forecasts in light of information [12]. These brain networks are propelled by the construction and capability of the human cerebrum, made out of various layers of interconnected hubs that cycle data. Through a process known as training, deep learning models learn to perform tasks by analyzing large datasets, identifying complex patterns, and making decisions or predictions without explicit programming [13].

2.3.2 ADVANTAGES IN HEALTHCARE

Deep learning offers numerous advantages in healthcare, revolutionizing various aspects of the medical field:

- **Enhanced Diagnostics:** Deep learning models exhibit high accuracy in interpreting medical imaging, aiding in early and accurate disease detection in radiology, pathology, and other diagnostic fields.
- **Predictive Analytics:** These models predict disease risks, patient outcomes, and treatment responses by analyzing complex datasets, empowering preventive care and personalized medicine.

Deep learning speeds up drug discovery by dissecting sub-atomic designs, anticipating drug associations, and smoothing out the medication improvement process. Modified treatment plans are contrived by thinking about individual patient information, advancing treatment viability, and limiting negative impacts. Automation of administrative tasks, improved hospital management, and enhanced resource allocation contribute to operational efficiency in healthcare settings.

2.4 APPLICATIONS OF DEEP LEARNING IN GENDER-BASED MEDICINE

Deep learning has found different and effective applications in orientation-based medication, utilizing its capacity to dissect complex datasets and uncover multifaceted examples well defined for various sexes [14]. In diagnostics, deep learning models have exhibited uncommon accuracy in deciphering clinical imaging information, supporting the early discovery and precision detection of orientation explicit medical issues as displayed in Figure 2.2. For instance, in breast cancer detection, these models have demonstrated promising results in analyzing mammograms and identifying subtle differences in tumors based on sex-specific characteristics.

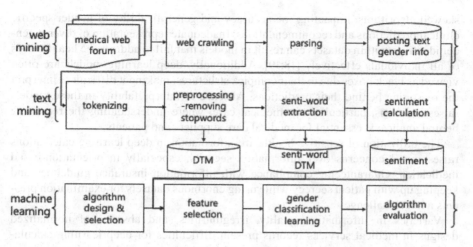

FIGURE 2.2 Machine learning in gender-based medicine.

Moreover, deep learning plays a crucial role in addressing gender disparities in mental health. By analyzing behavioral data and psychological records, these algorithms identify gender-specific patterns, symptoms, and responses to treatments. Such insights enable healthcare providers to develop personalized therapies and support strategies that address gender-specific nuances in mental health conditions. The applications of deep learning in gender-based medicine extend across diagnostics, predictive analysis, reproductive health, and mental health. Its ability to detect complex, gender-specific patterns enhances diagnostic accuracy, tailors treatment approaches, and deepens the understanding of gender-specific health issues, ultimately leading to improved healthcare outcomes and more equitable service delivery [15–17].

2.4.1 DIAGNOSIS AND DISEASE PREDICTION

Deep learning assumes a vital part in diagnosing orientation explicit medical issue and foreseeing illnesses by examining immense and different datasets. In orientation-based medication, these models display excellent precision in deciphering clinical imaging, like mammograms for breast cancer recognition or prostate X-rays for prostate malignant growth. They identify subtle gender-specific variations in disease presentation, aiding in early and accurate diagnostics. This enables timely interventions and enhances overall outcomes. In addition, deep learning empowers the distinguishing proof of orientation explicit biomarkers and examples in different wellbeing boundaries, working with the forecast of common or rare illnesses in different sexes.

2.4.2 SOME CHALLENGES AND LIMITATIONS ASSOCIATED WITH THE APPLICATION OF DEEP LEARNING IN GENDER-BASED MEDICINE

One of the primary challenges is the potential bias in datasets used to train deep learning models. If datasets are biased or inadequately diverse, they can lead to skewed algorithms, impacting the accuracy and generalizability of gender-specific clinical predictions and recommendations. Inadequate representation of diverse gender identities within datasets can result in models that fail to address the health needs of all individuals effectively [18,19]. Additionally, deep learning models are often viewed as black boxes due to their complex structures, making it difficult to interpret the reasoning behind their predictions. Absence of interpretability in these models raises concerns, particularly in medical care, where understanding the reasoning behind analyze is essential for medical care suppliers and patients.

The utilization of sensitive healthcare information in deep learning calculations raises moral concerns over information security, especially in orientation-based medication. Guaranteeing compliance with information insurance guidelines and keeping up with patient secrecy while using enormous datasets for examination presents a huge challenge.

Varieties in information quality, irregularities, and absence of normalized designs in medical services records present difficulties for deep learning calculations. Integrating diverse data sources and ensuring data quality and consistency across various healthcare systems and settings remain challenging, affecting the performance and reliability of the models.

Deep learning models developed on unambiguous datasets may need generalizability across various populations or medical care settings. The requirements for thorough approval and testing of these models across different populaces and medical services conditions to guarantee their strength and unwavering quality are not trivial.

Coordinating deep learning applications into clinical work processes and medical services frameworks presents difficulties concerning acknowledgment, convenience, and compatibility with existing practices. Resistance to adopting AI-driven solutions, concerns about reliability, and usability issues among healthcare professionals hinder the seamless integration of these technologies into clinical settings.

Addressing these difficulties and constraints in the utilization of deep learning in orientation-based medication requires cooperative endeavors from analysts, medical care experts, policymakers, and innovators Advancing diverse and representative datasets, ensuring model interpretability, adhering to ethical standards, improving data quality, validating models across varied populations, and integrating AI technologies into clinical workflows are crucial steps in harnessing the full potential of deep learning for gender-based medicine while addressing related challenges.

2.5 CASE STUDY: ENHANCING BREAST CANCER DIAGNOSIS USING DEEP LEARNING

In the domain of clinical diagnostics, the reconciliation of state of the art advances has altogether progressed our capacity to recognize and address wellbeing concerns. One such area of focus is the conclusion of breast cancer, a basic part of women's health. This contextual investigation delves into the use of deep learning methods to improve the accuracy and effectiveness of breast cancer diagnosis. Using a dataset from Kaggle, specifically the 'breast histopathology-images' collection, our study aims to explore the potential of artificial intelligence in transforming the field of pathology and enhancing the efficiency and timeliness of breast cancer detection.

This dataset, rich in histopathological images, serves as a valuable resource for training and validating deep learning models, ultimately leading to improved diagnostic outcomes in breast cancer. The dataset comprises a total of 20,000 histopathological images: 15,701 images without Intrusive Ductal Carcinoma (IDC) (IDC(–)) and 4,299 images with IDC present (IDC(+)). Each image has dimensions of 50 x 50 pixels and three color channels.

To guarantee consistency and similarity with AI models, the pixel values in the dataset are at first scaled inside the scope of 0 to 256. Notwithstanding, with the end goal of this investigation, a normalization step is applied, downsizing the pixel values to a range from 0 to 1. This scaling works with combination during model preparation and guarantees reliable information portrayal.

To evaluate the model's speculation execution, 20% of the dataset is saved for testing. This dividing assesses the model on inconspicuous information, giving bits of knowledge into its capacity to sum up past the preparation set. Tending to the test of imbalanced class measures, an undersampling system is carried out. This includes lessening the quantity of tests in the over-addressed class (IDC(–)) to adjust the class dispersion. By adopting this approach, the model is trained on a more balanced dataset, mitigating potential biases toward the majority class and enhancing its ability to accurately classify both IDC(+) and IDC(–) cases. The distribution of target classes is illustrated in Figure 2.3, which shows the number of occurrences for each class on the y-axis. This visualization clearly depicts the dataset imbalance, highlighting the greater prevalence of IDC(–) cases compared to IDC(+) cases.

The deep learning model accomplished a general exactness of 82.68%. The performance metrics for each class, IDC(–) and IDC(+), are as follows: For IDC(–), the accuracy is 84%, the recall is 81%, and the F1-score is 82%. For IDC(+), the accuracy is 82%, the recall is 84%, and the F1-score is 83%. The large scale normal and weighted-normal measurements, both at 83%, show a reasonable exhibition across classes, showing the model's viability in bosom malignant growth determination on the given dataset.

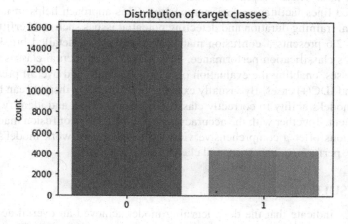

FIGURE 2.3 Distribution of target classes.

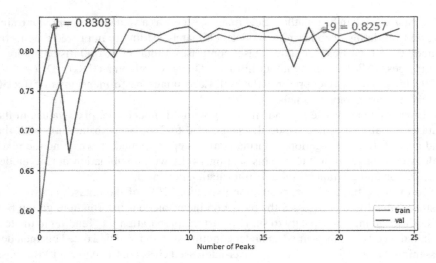

FIGURE 2.4   Evaluation function for ANN.

The assessment capability given fills in as a powerful device for surveying the presentation of a brain network model as displayed in Figure 2.4. The primary plot, known as the expectation to learn and adapt, gives a visual portrayal of the model's preparation and approval exactness and misfortune over ages. This diagram is pivotal for recognizing patterns in assembly and potential overfitting. The x-hub indicates the quantity of preparing ages, while the y-pivot outlines the exactness and misfortune values, offering experiences into the model's learning elements.

The one more arrangement of plots in Figure 2.5 offers a unique point of view on the precision and misfortune changes all through the preparation cycle. The visuals depict separate lines for training and validation accuracy, as well as training and validation loss. These graphs track the model's performance across epochs, with color-coded lines facilitating easy comparison. This approach helps in identifying the optimal training duration and detecting potential issues such as overfitting.

Figure 2.6 presents a confusion matrix that provides a detailed breakdown of the model's classification performance. This matrix aligns actual classes with predicted classes, enabling the evaluation of accuracy, recall, and overall precision for IDC(–) and IDC(+) cases. By visually examining this matrix, insights can be gained into the model's ability to correctly classify different classes and identify areas for improvement. Together with the accuracy, loss plots, and the confusion matrix, these visualizations offer a comprehensive view of the neural network model's training dynamics, performance metrics, and classification outcomes.

2.6   RESULT

The results indicate that the deep learning model achieved an overall accuracy of 82.68%. This figure represents the proportion of correctly classified instances out

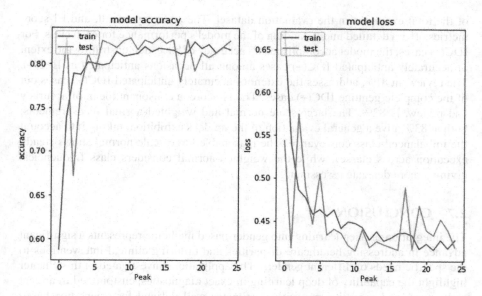

FIGURE 2.5 The accuracy and loss changes.

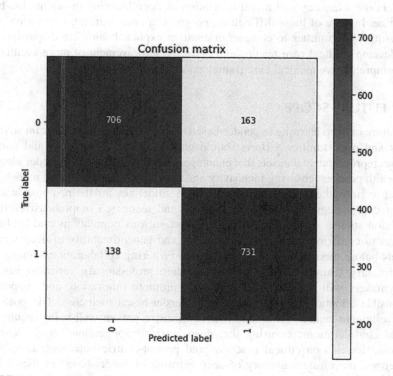

FIGURE 2.6 Confusion matrix.

of the total examples in the evaluation dataset. The accuracy, recall, and F1-score metrics offer a detailed understanding of the model's performance for each class. For IDC(–) cases, the model accomplished an accuracy of 84%, demonstrating the extent of accurately anticipated IDC(–) cases among all occasions anticipated as IDC(–). The review, at 81%, addresses the extent of accurately anticipated IDC(–) cases out of the complete genuine IDC(–) cases. The f1-score, a consonant mean of accuracy and review, is 82%. The large-scale normal and weighted-normal measurements, both at 83%, give a general evaluation of the model's exhibition, taking into account the imbalanced class conveyance. The reasonable large-scale normal shows steady execution across classes, while the weighted-normal considers class frequencies, giving a more delegate assessment.

2.7 CONCLUSION

The integration of deep learning into gender-based medicine represents a significant advance in addressing healthcare disparities and tailoring clinical interventions to the specific needs of different genders. The applications investigated in this chapter highlight the capability of deep learning in exact diagnostics, customized treatment techniques, and proactive preventive estimates well defined for orientation-based healthcare. In any case, difficulties, for example, information predisposition, interpretability, moral considerations, and clinical reception, should be addressed to augment the adequacy and moral utilization of deep learning in orientation-based medicine. In spite of these difficulties, progress in deep learning innovation offers promising opportunities to conquer orientation explicit healthcare disparities, further develop medical care results, and drive the improvement of more evenhanded and comprehensive medical care frameworks.

2.8 FUTURE SCOPE

The future of deep learning in gender-based medicine promises significant advancements and opportunities. Efforts should concentrate on developing and curating diverse, representative datasets that encompass a broad spectrum of gender identities and health profiles, ensuring inclusivity and accuracy in deep learning models. It is crucial to strengthen ethical frameworks and guidelines for the responsible use of patient data, safeguarding privacy, consent, and security. Comprehensive clinical validation studies should be conducted across various populations and healthcare settings to confirm the reliability, accuracy, and generalizability of deep learning models before their clinical implementation. Fostering collaborations among multidisciplinary teams of researchers, healthcare professionals, technologists, and policymakers will help address challenges, promote innovation, and support the responsible adoption of deep learning in gender-based medicine. The goal is to achieve human-centred AI solutions that prioritize patient well-being, equity, and ethical considerations, ensuring that deep learning applications in gender-based medicine focuses on ethical practices and patient-centric outcomes. Despite the challenges, the future trajectory of deep learning in gender-based medicine holds immense potential for transforming healthcare delivery, reducing gender-specific health disparities, and advancing toward more personalized, equitable, and inclusive healthcare systems. Ongoing research, development, ethical considerations, and collaborative efforts will drive the impactful role of deep learning in gender-based medicine in the coming years.

REFERENCES

[1] Iqbal, K., & Majeed, H. A. (2023). From prophet Muhammad's era to modern times: Empowering women in medical activities for a healthier Pakistan. *INKISHAF, 3*(8), 116–133.

[2] Wang, J., Wang, J., Wang, S., & Zhang, Y. (2023). Deep learning in pediatric neuroimaging. *Displays, 80,* 102583.

[3] Munive, A., Donville, J., & Darmstadt, G. L. (2023). Public leadership for gender equality: A framework and capacity development approach for gender transformative policy change. *Eclinicalmedicine, 56.*

[4] Sihare, S. R. (2024). Student dropout analysis in higher education and retention by artificial intelligence and machine learning. *SN Computer Science, 5*(2), 202.

[5] Sarker, F., Mamun, K. A., Uddin, J., Ahmed, S., Nisha, N. A., & Joshi, D. (2023). Digital inclusion in agri-food systems in Bangladesh: The digital innovation and transformation initiative. *Journal of Digital Agriculture, 12*(4), 234–245.

[6] Rupert, D., Thompson, B., Fernando, M., Kays, M., Barry, P., & Jain, S. (2021). Proceedings of the third annual women in medicine summit: An evolution of empowerment 2021. *International Journal of Academic Medicine, 7*(4), 307–357.

[7] Wang, F. (2023). *Deep Learning Applications in Bone Mineral Density Estimation, Spine Vertebra Detection, and Liver Tumor Segmentation* (Doctoral dissertation, University of Maryland, College Park).

[8] Siddiqui, S. S., Loganathan, S., Elangovan, V. R., & Ali, M. Y. (2023). Artificial intelligence in precision medicine. In *A Handbook of Artificial Intelligence in Drug Delivery,* 531–569. Academic Press.

[9] Osman, M., & Elamin, I. (2023). Advancing ethical and sustainable economy: Islamic finance solutions for environmental, social, & economic challenges in the digital age. *International Journal, 10*(5), 408–429.

[10] Benck, K. N., Khan, F. A., & Munagala, M. R. (2022). Women in mechanical circulatory support: She persisted! *Frontiersin Cardiovascular Medicine, 9,* 961404.

[11] Heifetz, I., & Jaffe, P. G. (2023). Exploring the impact of industrialization on social mobility in rural communities: Towards inclusive and sustainable economic transformation. *Law and Economics, 17*(3), 218–236.

[12] Campbell, N. D., Ettorre, E., Campbell, N. D., & Ettorre, E. (2011). Conclusion: Making gender matter in an age of neurochemical selves. In *Gendering Addiction: The Politics of Drug Treatment in a Neurochemical World,* 182–202.

[13] Pantelic, J., Nazarian, N., Miller, C., Meggers, F., Lee, J. K. W., & Licina, D. (2022). Transformational IoT sensing for air pollution and thermal exposures. *Frontiersin Built Environment, 8,* 236.

[14] Arora, R., Mutz, D., & Mohanraj, P. (Eds.). (2023). *Innovating for the Circular Economy: Driving Sustainable Transformation.* CRC Press.

[15] Wade, B., Bathrooms, C. E., & Street, W. (2020). Transforming the toilet. In *Bathroom Battlegrounds: How Public Restrooms Shape the Gender Order,* 162–178. Routledge.

[16] Arner, D. W., Animashaun, S., Charamba, K., & Cai, Y. (2022). MSME access to finance: The role of digital payments. *United Nations, Economic and Social Commission for Asia and the Pacific, MSME Financing Series*, 7.

[17] Jin, X., Kim, E., Kim, K. C., & Chen, S. (2023). Innovative knowledge generation: Exploring trends in the use of early childhood education apps in Chinese families. *Journal of the Knowledge Economy*, 1–40.

[18] Kaur, G., Gupta, M., & Kumar, R. (2021). IoT based smart healthcare monitoring system: A systematic review. *Annals of the Romanian Society for Cell Biology*, 3721–3728.

[19] Gupta, M., Ahmed, S., Kumar, R., & Altrjman, C. (Eds.). (2023). *Computational Intelligence in Healthcare: Applications,Challenges, and Management*. CRC Press.

3 From Data to Diagnosis
AI's Role in Gender-Responsive Care

Thamizhiniyan Natarajan and
Shanmugavadivu Pichai

3.1 INTRODUCTION

Over the past few years, artificial intelligence (AI) has been a rapidly emerging technology in healthcare, disrupting various aspects related to medical diagnosis, treatment, and patient care [1]. Inclusion of AI technologies may carry enormous potential for higher efficiency, precision, and access in service delivery in healthcare across the globe [2]. From predictive analytics to offering personalized treatment recommendations, AI will radically reshape the health landscape [3].

Unique health needs of diverse demographic groups, including those based on gender, should also be acknowledged within this progression. Disparities and biases that were likely to have occurred or that exist in the diagnosis and treatment of health conditions between the two genders in the past normally were disregarded or handled inappropriately by healthcare systems [4]. This has led to some populations, particularly women and individuals identifying as non-binary or other gender minorities, suffering disproportionately in accessing quality healthcare services [5].

With this background, the concept of gender-responsive care is emphasized, aiming to reduce these disparities so that every individual, regardless of gender orientation, receives equitable outcomes from healthcare. Gender-responsive care is rather holistic and is based on the diverse health needs and experiences of different individuals in terms of their gender and class in tailoring medical interventions and services accordingly. Gender mainstreaming in healthcare practices helps providers address the unique risk factors, symptoms, and treatment responses associated with different health conditions [6].

More specifically, this chapter seeks to examine AI and gender-responsive care in health. The chapter's specific objectives are to:

- **Examine the Role of AI:** Analyze how AI technologies are contributing to gender-responsive care from data collection to diagnostic procedures.
- **Identify Challenges and Opportunities:** Discuss the challenges and opportunities in applying AI to health systems in a way that will allow the scaling up of gender equity and reducing disparities.

DOI: 10.1201/9781003473435-3

- **Use Case Examples to Highlight Best Practices:** Use cases to highlight best practices in gender-responsive care using AI that have been successfully applied and informed by real-world case studies and research findings.

This chapter will contribute to the growing need for the optimization of health delivery through inclusive, patient-centered approaches in light of the synergies that exist between AI and gender-responsive care. In reality, the needs of diverse people throughout societies need to be set as a priority across the gender spectrum.

3.2 UNDERSTANDING GENDER-RESPONSIVE CARE

Healthcare should recognize and meet the health needs and experiences of individuals based on their gender identity. Then the practice is said to be gender-sensitive healthcare. To achieve a gender-sensitive healthcare, doctors and clinical experts should provide tailored treatments, interventions and support services that consider social and cultural factors that influence health outcomes across various genders. It is vital to acknowledge gender differences in healthcare to make it inclusive, equitable and responsive to the needs and identities of all people in the society.

It is very important to consider the unique needs of an individual's gender when providing healthcare. Healthcare should take care of gender identities and, in turn, that will promote equality and inclusivity by addressing the disparities in gender biases, stereotypes and systemic inequalities. When healthcare professionals recognize and address the concerns of genders, they can deliver more personalized and effective treatment that might lead to enhanced patient-care outcomes and provide a better quality of life for individuals across the gender spectrum. Gender-responsive healthcare focuses on tailored interventions and early prevention targeted at the risk factors that are associated with gender. This strategy will reduce the impact of diseases and will promote the overall wellness of individuals.

The impact of gender differences on the health outcomes should be thoroughly studied while developing healthcare strategies. The biological variations between males and females can influence how they will respond to treatments. Societal norms linked to gender play a significant role in shaping behaviours related to healthcare. These gender-related behaviours require a strategy that will take into account the interconnected effects of biology, culture and social factors in health.

3.2.1 CHALLENGES IN CURRENT HEALTHCARE

Despite the growing recognition of the importance of delivering gender-sensitive care, healthcare systems worldwide still face many challenges in providing solutions that account for gender differences. An important challenge is the existence of prejudice that is ingrained in the healthcare professionals that can ultimately influence treatment and diagnostic decisions. These biases will lead to disparities in the quality of care provided to individuals and may contradict the principles of inclusivity in healthcare [4].

Another challenge is the availability of data with gender-specific features. There is also very little research on gender-specific health issues. Without the availability of robust data, it is challenging to make evidence-based decisions and interventions tailored to individuals' health conditions [7][8]. It is also evident that healthcare professionals may lack adequate knowledge or training in gender-sensitive communication and provision of inclusive care. This might further worsen the already existing disparities in healthcare delivery [9].

Lack of policy frameworks and health infrastructure that addresses the specific needs of different genders can pose a challenge to gender-responsive services and support. Issues like gender-based discrimination, social stigma and limited number of gender-sensitive healthcare facilities can deny communities from receiving timely and appropriate healthcare [10]. Increasing awareness and integrating gender considerations into medical education and clinical practice can address these challenges. Empowering patients to advocate for these gender-specific needs and preferences within the healthcare setting will also help mitigate the challenges.

3.3 DATA COLLECTION AND AI IN HEALTHCARE

Accurate data collection in healthcare is critical. If there are biases in healthcare data, it can have an impact on the diagnoses and treatment outcomes. In general, biases in in data collection may arise due to disparities in representation of populations and overlooking of nuanced health experiences [11]. For example, in research and clinical trials, biases have led to some demographic groups not being represented properly, which ultimately resulted in incomplete and potentially biased datasets [12].

Machine learning models are significantly influenced by data that are biased. These models will produce biased results that would worsen the inequalities in healthcare. There is also a concern that the biases present in the data may undermine trust in the system and increase disparities in healthcare access [13]. This could ultimately prevent marginalized groups from receiving adequate healthcare benefits.

There needs to be a strategy to address biases in healthcare data. The strategy should include collecting data that encompasses demographic information, socio-economic factors and cultural differences. It should also employ data collection techniques to identify and remove the biases during all the stages. Collaboration between healthcare experts, researchers, policy makers and community stakeholders play a vital role in promoting equity and inclusivity in data collection efforts [14].

3.3.1 AI IN HEALTHCARE

The surge of AI in diverse domains including healthcare has tremendously sent the art of computing in a new trajectory, assuring improved precision in prediction and classification. The AI models developed under the banner of machine learning and deep learning have revolutionized the wide array of domains including natural language processing and computer vision. The application of AI in such domains has opened new vistas in early diagnosis, detection, classification of diseases without human intervention [15].

The availability of voluminous medical data and medical images has redefined the power of computing with the support of AI and advanced computing hardware. The AI models are empowered to harness the hidden patterns among the attributes of the medical datasets and accordingly formulate mapping functions between the input attributes and target attributes. The roles of AI algorithms in temporal analyses of medical data promise to surpass traditional computing models and human diagnosis [16].

The potential of AI in early prediction of certain diseases like breast cancer, lung cancer, brain tumours, Parkinson's disease and cardiovascular diseases is proven to be a boon for medical practitioners as well. The proven success of disease diagnosis using AI is recognized as one of the most reliable medical procedures, and hence the development of AI-based models for the healthcare domain has triggered the growth of AI-based medical diagnostics.

3.4   GENDER-SPECIFIC CONSIDERATIONS IN AI

The automation of computerized healthcare has triggered the meticulous compilation of health information in the form of electronic heath records. These health records contain multiple attributes which directly or indirectly contribute to the process of disease diagnosis and detection. The difficulty in understanding the relationship between disease attributes and predictive factors limits data exploration to manual processes. In this context, it is imperative that AI algorithms are capable of discovering the inherent patterns against even hundreds of input variables effortlessly and formulating the relationship between input attributes and output attributes. It is important to emphasize that the accuracy of prediction and classification of AI models ideally anchor on the distribution and diversity of data depicting the characteristics of disease and their correlation to the disease under investigation.

The diversity and coherence of data despite its volume plays a very important role in the generalization of an AI model. Hence researchers indulging in development models in healthcare solutions are equally focusing on data curation and data imputation in order create intact datasets. By and large it is observed that the publicly available healthcare datasets are to be necessarily subjected to exploratory data analysis (EDA) in order to unveil the data discrepancies such as missing values, outliers and irrelevant data. Hence the researchers have paramount responsibility of curating the data prior to the development of AI models. In addition, certain datasets must be subjected to data normalization and encoding as per the requirements of AI models. Hence standardization of datasets demands equal amount of attention and importance as that of development of AI models for healthcare applications. Moreover, the other common problems posed by the medical datasets are bias and variance. Bias in datasets is characterized by the inaccurate representation of population of the dataset. Variance refers to the exorbitant variation of data points against the mean average of the population. These problems can be suitably mitigated by adopting suitable pre-processing methods.

The medical world is witnessing new trends in the spread of gender-specific diseases among both genders. For instance, breast cancer which was considered as female-centric is now being diagnosed among men also. The fast-paced lifestyle and life management has vanished the barriers of gender-specific disease thereby calling for creation and construction of new datasets meeting the requirements of new trends in healthcare diagnostics and therapeutics. Hence the researchers are driven to generate gender-neutral datasets in order to assure generalization in the AI models and consistent accuracy in prediction and classification of diseases.

3.5 CASE STUDIES AND EXAMPLES

This section is designed to outline six distinct applications of AI models in health-care. These case studies assume relevance and importance in exemplifying the power of AI in providing solutions to gender related health concerns.

3.5.1 AI-POWERED BREAST CANCER SCREENING

The diagnosis of breast cancer at an early stage through medical image analysis remains a challenge for radiologists and oncologists due to the limitations of the manual diagnosis using the medical imaging modalities such as X-ray and mammogram. However, AI-based breast cancer diagnosis models are proven to be highly effective and precise in early detection and diagnosis of breast cancer. This has motivated medical practitioners including oncologists to use such models for screening, treatment plan and follow ups. It is apparent that the collaboration between Google Health and Imperial College London for the development of AI-based algorithms for breast cancer diagnosis have been highly accurate and reliable as endorsed by medical experts. Such research outcomes will help in greater ways in taking the technology to under-privileged society benefiting rural women and there by combating this fatal disease at an early stage [17].

3.5.2 AI-BASED MENTAL HEALTH SCREENING

The emergence of newer dimensions in cognitive psychology and cognitive science has promoted the application of AI in monitoring human mental and behavioural activity. Diagnosis of mental health issues through human intervention is deemed too cumbersome due to the fact that the cognitive functioning and response systems radically vary from one person to another. Hence developing a generic system addressing these practical challenges requires sophisticated technology enabled solution as addressed by AI models. The demands for such models have promoted the development of newer and innovative mobile apps for mental health screening and monitoring. Woebot is an illustrative chatbot that is designed for mental health screening and monitoring based on the principle of natural language processing [18]. This chatbot is capable of drawing inferences from the cognitive behaviour of an individual and then recommending suitable therapeutic solutions. Due to the educational, social, economic and cultural implications of gender equity, women may be deprived of gaining equal access to remedies for mental health–related issues. Availability of such chatbots empower women to break this barrier and gain equal access through technology driven diagnosis and therapy.

3.5.3 PREDICTIVE ANALYTICS FOR CARDIOVASCULAR DISEASE

The accurate diagnosis of cardiovascular disease (CVD) requires drawing observations on diet plan, physical activity, biochemical analysis of blood and functioning of heart and blood vessels. Development of powerful AI-driven CVD diagnosis and detection by eliciting holistic inferences from all these factors will serve as a supplementary system for cardiologists. Access to longitudinal data on patients at risk for CVD is crucial for advancing diagnosis and research. As an experimental effort to delineate the primary risk factors of CVD, the Framingham Heart Study [19] has been committed to collecting data on epidemiology of coronary heart disease (CHD) since 1948. This ground-breaking availability of data on CHD has enabled the development of AI powered algorithms for accurate diagnosis of CVD. The statistics on CVD is noted as one of the leading causes of morbidity and mortality in men. However, recent studies report that CVD takes a toll on women also due to lack of timely prognosis and diagnosis. This scenario calls for development of generic AI-based CVD diagnosis models for both men and women.

3.5.4 AI-Assisted Reproductive Health Services

In developing countries, women's health issues from puberty to menopause have not received adequate attention and care, leading to pronounced gender discrimination that negatively impacts the overall well-being of women and girls. To align with the objectives of United Nation in promoting and sustaining women's health, the Ava bracelet has been introduced. Ava is an innovative wearable device powered by pack of AI algorithms designed to continuously monitor and record vital parameters on women's menstrual cycles and thereby precisely predict the potential fertility window for conception. This device records the physio-psychological factors – namely skin temperature, heart rate fluctuations and sleep cycles – of prospective women planning for conception [20] [21]. Needless to mention the health factors of women during pregnancy have a direct correlation with the health of newborns, leading to healthy future generation.

3.5.5 Personalized Medicine

The application of AI has increased by leaps and bounds beyond disease diagnosis and detection in healthcare. The ongoing AI research in pharma has unfolded a new domain of research, namely personalized medicine. The new developments in the domain of personalized medicine find a place in precise patient-specific treatment patterns based on the respective attributes of health ailments or disorders. This ground-breaking AI research in personalized medicine helps physicians customize treatment pattern focusing more on cause-effect of the disease which varies among gender, age, weight, demography and other factors.

3.5.6 Virtual Health Assistants

The reliability of AI models in healthcare services has led to the development of virtual health assistants. These AI models can help provide primary medical assistance in remotely located villages where availability of health workers and medical practitioners is sparse. Ada is a typical virtual health assistant capable of offering consultation to patients from primitive to complex healthcare consultation and services. Such assistants provide healthcare services that are free from biases and gender preferences.

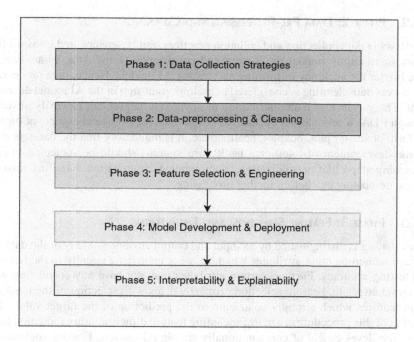

FIGURE 3.1 AI-powered life cycle.

3.6 AI-POWERED LIFE CYCLE: FROM DATA COLLECTION TO DIAGNOSIS

The development life cycle of an AI model for any given application domain involves a generic sequence of five steps, spanning from data collection to deployment. Each step assumes an equal quotient of significance on the quality assurance of the AI-models. Figure 3.1 portrays those distinct steps, and the purpose of each step is described in this section.

3.6.1 PHASE 1: DATA COLLECTION STRATEGIES

The first step in the life cycle of AI in healthcare deals with the problem-specific data collection from personal health records (PHR), electronic health records (EHR) and wearable devices, etc. Since data diversity and distribution are crucial, preparing healthcare datasets requires significant effort to ensure they accurately and proportionately represent the population, considering factors such as age, gender, and demographics. Standardizing datasets with gold standards and benchmarks is one of the most important steps in this process. Ethical and legal concerns levy certain restrictions on dataset preparation, such as anonymization. Data protection and privacy are also major concerns when handling sensitive datasets.

3.6.2 PHASE 2: DATA PRE-PROCESSING AND CLEANING

The flaws in data collection and collation practices, faulty sensors, and transmission errors significantly impact the quality and consistency of the data, which can, in turn, hinder the accuracy of training and testing AI models. Hence, data processing that covers data cleaning is considered as an important step in the AI model development. The variance in data distribution of any given dataset can be easily unveiled through EDA. It provides an exhaustive insight into the statistical aspects of data in the form of scatter plot, boxplot, heatmap etc. It is mandatory that the datasets used for the development of intelligent healthcare systems should be subjected to pre-processing steps like normalization, discretization, aggregation, handling missing values, de-noising etc. before further processing.

3.6.3 PHASE 3: FEATURE SELECTION AND ENGINEERING

Every dataset is distinguished by its input and output attributes. Many of the datasets contain numerous input attributes which may not contribute equally to the training and testing accuracy. Further, certain attributes may not have any correlation with the target attribute. Feature selection primarily deals with selection of the most relevant features which strongly contribute to the prediction of the target value. The benefits of this procedure are many, including feature dimensionality reduction leading to the development of computationally simple AI models. Feature engineering deals with augmenting the feature sets of a dataset by combining rationally one or more related attributes. Researchers are expected to apply feature selection and engineering strategies as per the objective of the problem domain.

3.6.4 PHASE 4: MODEL DEVELOPMENT AND DEPLOYMENT

Model development consists of model selection, model training and model deployment. Model selection involves selection of the right type of AI model that solves the problem at hand. There are a wide variety of algorithms that are available and AI practitioners should carefully select the appropriate algorithms that will give the highest measures of results. Common models that can be considered are neural networks, decision trees, logistic regression, ensemble methods etc. Effort needs to be spent on the selection of tools like TensorFlow, PyTorch and Scikit-learn which can help the practitioners in efficient development and implementation of the AI model. After the selection of the right kind of model, the next step is to train the model. This involves adjusting the model's parameters and evaluating its performance until the desired accuracy is achieved. After training the model, the next step is deploying it in a production environment. In model deployment, the model is integrated into the medical diagnostic process, and it must be ensured that the model is handling real-time data efficiently.

3.6.5 PHASE 5: INTERPRETABILITY AND EXPLAINABILITY

AI models are generally considered to be black boxes. There is no visibility into the internal details of the exact operations that the model is executing to arrive at the results of prediction or classification. Due to the black box nature of the AI models, the medical practitioners may hesitate in using the AI models in predictions of diseases. They find it hard to completely believe a model's recommendation that functions like a black box. In the healthcare domain it is critical to know the reason behind the decisions taken by the AI model. Explainable AI (XAI) is a set of tools and frameworks that helps the stakeholders in understanding how the AI models function internally to arrive at a decision. XAI provides insights into the functioning of the model, and it brings in transparency on how the decision is being made. XAI helps increase the much-needed trust in AI model's decision-making capability.

The emerging challenges and opportunities in gender-based healthcare should be explored and expanded with due consideration in every phase of AI model development from data collection to disease diagnosis. These concerns are expected to open up new dimensions in the development of AI-based solutions in gender-based disease diagnosis.

3.7   THE GENDER GAP IN HEALTH DATASETS: A CASE STUDY

Gender information is an important variable in healthcare datasets. The information pertaining to gender helps in developing AI driven gender-sensitive healthcare. Many datasets lack gender information that can significantly affect the decision-making process in providing diagnoses and treatments. This section presents an experiment highlighting the absence of gender data in the health dataset and how it impacts tailored healthcare for gender. The goal of this study is to understand the gap in gender information in some of the health datasets. This case study demonstrates the challenges faced by AI models in delivering diagnostic results when there is a lack of data related to gender.

Table 3.1 provides an overview of the selected health datasets, including information, on the number of cases and how female and male subjects are distributed.

By examining the collected health datasets, it can be inferred that there is a disparity in how the gender-related information is included. Only certain datasets like those focusing on lung and breast cancer provide gender-specific data. But the majority of healthcare datasets pertaining to heart diseases and brain tumours do not contain gender-specific data. These inconsistencies shed light on the challenges faced by AI algorithms in delivering results without gender-sensitive healthcare outcomes.

3.7.1   GENDER DISTRIBUTION ANALYSIS ACROSS HEALTH DATASETS

Table 3.2 provides an analysis of the datasets to show how gender data is distributed. It also reveals variations in healthcare results based on gender.

These graphs highlight the importance of considering gender data in training and validation of AI algorithms. For the AI algorithms to perform well in gender-sensitive healthcare, the dataset should contain data pertaining to all genders.

TABLE 3.1
Gender distribution in publicly available health datasets.

S. No.	Dataset	Source	No. of Instances	No. of Female Subjects	No. of Male Subjects
1.	Lung Cancer	www.kaggle.com/datasets/thedevastator/cancer-patients-and-air-pollution-a-new-link	1000	402	598
2.	Lung Cancer	www.kaggle.com/datasets/raddar/smoking-related-lung-cancers	53427	31522	21905
3.	Lung Cancer	www.kaggle.com/datasets/mysarahmadbhat/lung-cancer	284	147	137
4.	Cardiovascular Disease	www.kaggle.com/datasets/sulianova/cardiovascular-disease-dataset	70000		
5.	Heart Disease	https://github.com/kb22/Heart-Disease-Prediction/blob/master/dataset.csv	303 (50,143)		
6.	Covid effect on Liver Cancer Prediction	www.kaggle.com/datasets/fedesoriano/covid19-effect-on-liver-cancer-prediction-dataset	450	333	117
7.	BRCA	www.kaggle.com/datasets/amandam1/breastcancerdataset	341	330	11
8.	Breast Cancer Wisconsin	www.kaggle.com/datasets/uciml/breast-cancer-wisconsin-data	546		
9.	Breast Cancer	https://datahub.io/machine-learning/breast-cancer#resource-breast-cancer	286		
10.	Breast Cancer	https://data.world/datagov-uk/5624cab2-85cc-4932-bb6e-63d67e49f516/workspace/file?filename=14-day-cancer-waiting-times-3.csv	673		
11.	Brain Tumor Dataset	www.kaggle.com/datasets/preetviradiya/brian-tumor-dataset	4600		

3.8 CONCLUSION

This chapter explores how AI systems can perform to their optimum level when gender-specific information included in very stage of the AI lifecycle. It is highlighted that gender-sensitive information is vital to ensure a holistic healthcare initiative. Moreover, ethical and privacy concerns are also addressed when the AI technologies are integrated into the healthcare systems. The chapter also discussed the importance of data collection methods, biases in algorithms and safeguarding the patient privacy. The chapter outlined the AI lifecycle stages from data collection to making diagnoses. Case studies and examples of AI technologies being used in healthcare highlighted the importance of gender-sensitive treatments.

TABLE 3.2
Gender Distribution Analysis

Gender Distribution	Analysis
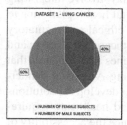	The graph pertaining to Dataset 1: Lung Cancer has 40% men and 60% women. This indicates that there is a higher incidence of lung cancer among men. AI algorithms can use this data to identify risk factors that are prevalent in men. Healthcare professionals can provide customize their treatment that can take in to account the smoking or other lifestyle habits among the men.
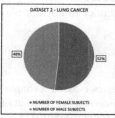 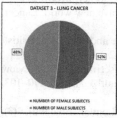	The Lung Cancer datasets 2 and 3 has evenly distributed male and female population. The lung cancer affects both the genders in the population from this demography. AI algorithms can use this insight to make sure the decision-making process is unbiased towards any gender.
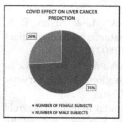	The dataset that contains the COVID-affected patients with cancer indicated a disparity in the gender distribution with 74% male and 26% female from the population. This could imply that the male population is affected more than the female population. This could be because of the lifestyle habits like smoking and alcohol consumption being prevalent more in men rather than in women. AI systems can employ intelligence to monitor predictions that can consider the gender disparity in this dataset.
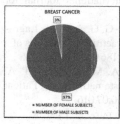	It can be inferred from the graph related to the breast cancer that 97% of women compared to 3% of men are affect by the disease. This dataset can be used to raise awareness that breast cancer can also affect men and AI systems can help in preparing a process to identify the disease in early stages for men.

3.8.1 FUTURE DIRECTIONS

The future of AI's role in gender-specific healthcare is very bright. There are lot of opportunities to explore the improvements that can be made to accuracy, speed and personalized treatments for all genders. Future progress may involve integrating AI with fields such as genomics, precision medicine and digital health platforms to provide healthcare solutions. It is important to strengthen the ethical and regulatory frameworks designed specifically for gender responsive healthcare. AI experts and healthcare professionals should collaborate to establish guidelines and standards that prioritize transparency, accountability and fairness when using AI driven gender-sensitive healthcare. Future research initiatives should aim to develop AI solutions that prioritize the needs of patients and cater to gender-based healthcare requirements. Future progress should leverage AI technology to enhance the accessibility of healthcare services for marginalized groups such as women, gender minorities and other vulnerable populations.

REFERENCES

[1] Lee, D., & Yoon, S. N. (2021). Application of artificial intelligence-based technologies in the healthcare industry: Opportunities and challenges. International Journal of Environmental Research and Public Health, 18(1), 271. doi: 10.3390/ijerph18010271

[2] The Lancet. (2019). Artificial intelligence in global health: A brave new world. Lancet, 393(10180), 1478. doi:10.1016/s0140-6736(19)30814-1

[3] Dave, M., & Patel, N. (2023). Artificial intelligence in healthcare and education. British Dental Journal, 234(10), 761–764. doi:10.1038/s41415-023-5845-2

[4] Govender, V., & Penn-Kekana, L. (2008). Gender biases and discrimination: A review of health care interpersonal interactions. Global Public Health, 3(sup1), 90–103. doi:10.1080/17441690801892208

[5] Hamberg, K. (2008). Gender bias in medicine. Women's Health (London, England), 4(3), 237–243. doi:10.2217/17455057.4.3.237

[6] Celik, H., Lagro-Janssen, T. A. L. M., Widdershoven, G. G. A. M., & Abma, T. A. (2011). Bringing gender sensitivity into healthcare practice: A systematic review. Patient Education and Counseling, 84(2), 143–149. doi:10.1016/j.pec.2010.07.016

[7] Weber, A. M., Gupta, R., Abdalla, S., Cislaghi, B., Meausoone, V., & Darmstadt, G. L. (2021). Gender-related data missingness, imbalance and bias in global health surveys. BMJ Global Health, 6(11), e007405. doi:10.1136/bmjgh-2021-007405

[8] Gahagan, J., Gray, K., & Whynacht, A. (2015). Sex and gender matter in health research: Addressing health inequities in health research reporting. International Journal for Equity in Health, 14(1). doi:10.1186/s12939-015-0144-4

[9] Celik, H. H., Klinge, I. I., van der Weijden, T. T., Widdershoven, G. G. A. M., & Lagro-Janssen, T. A. L. M. (2008). Gender sensitivity among general practitioners: Results of a training programme. BMC Medical Education, 8(1), 36. doi:10.1186/1472-6920-8-36

[10] Celik, H., Lagro-Janssen, T. A. L. M., Widdershoven, G. G. A. M., & Abma, T. A. (2011). Bringing gender sensitivity into healthcare practice: A systematic review. Patient Education and Counseling, 84(2), 143–149. doi:10.1016/j.pec.2010.07.016

[11] Celi, L. A., Cellini, J., Charpignon, M.-L., Dee, E. C., Dernoncourt, F., Eber, R., . . . for MIT Critical Data. (2022). Sources of bias in artificial intelligence that perpetuate healthcare disparities—a global review. PLoS Digital Health, 1(3), e0000022. doi:10.1371/journal.pdig.0000022

[12] Gianfrancesco, M. A., Tamang, S., Yazdany, J., & Schmajuk, G. (2018). Potential biases in machine learning algorithms using electronic health record data. JAMA Internal Medicine, 178(11), 1544. doi:10.1001/jamainternmed.2018.3763

[13] Rajkomar, A., Hardt, M., Howell, M. D., Corrado, G., & Chin, M. H. (2018). Ensuring fairness in machine learning to advance health equity. Annals of Internal Medicine, 169(12), 866. doi:10.7326/m18-1990

[14] Arora, A., Alderman, J. E., Palmer, J., Ganapathi, S., Laws, E., McCradden, M. D., . . . Liu, X. (2023). The value of standards for health datasets in artificial intelligence-based applications. Nature Medicine. doi:10.1038/s41591-023-02608-w

[15] Davenport, T., & Kalakota, R. (2019). The potential for artificial intelligence in healthcare. Future Healthcare Journal, 6(2), 94–98. doi:10.7861/futurehosp.6-2-94

[16] Bajwa, J., Munir, U., Nori, A., & Williams, B. (2021). Artificial intelligence in healthcare: Transforming the practice of medicine. Future Healthcare Journal, 8(2), e188–e194. doi:10.7861/fhj.2021-0095.

[17] O'Hare, R. (2020, January 1). Artificial intelligence could help to spot breast cancer. Retrieved February 22, 2024, from Imperial News website: www.imperial.ac.uk/news/194506/artificial-intelligence-could-help-spot-breast/

[18] Relational Agent for Mental Health. (2021, November 9). Retrieved February 22, 2024, from Woebot Health website: https://woebothealth.com

[19] O'Donnell, C. J., & Elosua, R. (2008). Cardiovascular risk factors. Insights from Framingham heart study. Revista Espanola de Cardiologia (English Ed.), 61(3), 299–310.

[20] Goodale, B. M., Shilaih, M., Falco, L., Dammeier, F., Hamvas, G., & Leeners, B. (2019). Wearable sensors reveal menses-driven changes in physiology and enable prediction of the fertile window: Observational study. Journal of Medical Internet Research, 21(4), e13404. doi:10.2196/13404.

[21] Gupta, M., Ahmed, S., Kumar, R., & Altrjman, C. (Eds.). (2023). Computational Intelligence in Healthcare: Applications, Challenges, and Management. CRC Press.

4 Technology-Driven Approaches to Gender-Inclusive Healthcare

Sonam Lata and Priya Dubey

4.1 INTRODUCTION

Optimal progress within the field of biomedical engineering relies on staying abreast of the latest research trends and transitions. Breast cancer is one of the most common cancers, and is now the leading cause of death, in women around the world. According to the American Cancer Society's most recent findings, over 40,000 women and approximately 600 men died from breast cancer [1]. There are generally four types of breast cancer: benign, normal, in situ carcinoma, and invasive carcinoma [2, 3]. A tumor that is benign modifies the anatomy within the breast, but it cannot be considered toxic and does not meet the criteria for potentially hazardous cancer [4]. In contrast, in situ carcinoma impacts only the mammary duct lobules and fails to spread to adjacent organs [5]. This particular kind of cancer is not especially harmful and is curable if recognized early. Invasive carcinoma is the most severe form of breast cancer, with the possibility for progression to all other parts of the body [6]. Breast cancer has been confirmed using a variety of approaches over the years, including mammography, X-ray, ultrasound (US), positron emission tomography (PET), computed tomography, temperature measurement, and magnetic resonance imaging (MRI) [7, 8]. In this chapter, we will delve into the dynamic landscape of biomedical engineering, exploring the evolving research trends and their impact on the fields of breast cancer. Our investigation will thoroughly examine diverse scientific databases of breast cancer specifically, allowing us to identify and analyse pertinent studies using tools such as SPSS that align with specific focus on breast cancer. The outcomes of our research reveal the current state of biomedical engineering research, highlighting emerging trends and transitions. Notably, we observe the integration of novel technologies and methodologies based on transfer learning, including the transformative role of deep learning in breast cancer detection, presenting a paradigm shift in how data is analysed and interpreted. This chapter examines these technological advancements and discusses the challenges they pose, offering insights into potential solutions. By providing a nuanced exploration of the latest trends and transitions, this chapter offers valuable insights into the ever-evolving domain of breast cancer detection, specifically focusing on the transformative influence of deep learning, inspiring further innovative research in the field of breast cancer detection. Figure 4.1 shows the block diagram for the entire research methodology.

DOI: 10.1201/9781003473435-4

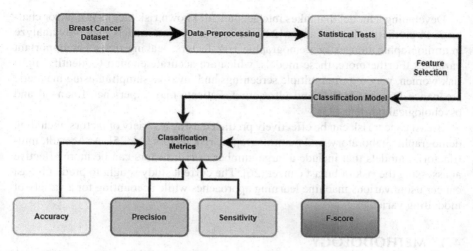

FIGURE 4.1   Block diagram to represent the proposed methodology.

4.2 RELATED WORK

Breast cancer is considered a multifactorial disease and the most common cancer in women worldwide [20], accounting for roughly 30% of all female cancers [9]. This disease has grown in prevalence over the last three decades, while death rates have decreased. However, mammography screening is estimated to reduce mortality by 20% while improving cancer treatment by 60% [10]. Diagnostic mammography can detect abnormal breast cancer tissue in patients who have subtle and inconspicuous malignancy signs. Because of the large number of images, this method cannot be used to assess cancer-suspected areas effectively. According to one study, approximately 50% of breast cancers were missed in screenings of women with dense breast tissue [11]. However, approximately one-quarter of women with breast cancer receive a negative diagnosis within two years of screening. As a result, early and accurate diagnosis of breast cancer is critical. The majority of mammography-based breast cancer screenings are performed on a regular basis for all women, typically once a year or every two years. A fix screening program for everyone is ineffective in diagnosing cancer at the individual level and may reduce the effectiveness of screening programs [12]. Experts, on the other hand, believe that combining mammography screening with other risk factors can lead to a more accurate diagnosis of at-risk women [13]. Furthermore, effective risk prediction through modelling can assist radiologists in setting up a personal screening for patients and encouraging them to participate in the early detection program, as well as identifying high risk patients [14].

Machine learning, as a modelling approach, refers to the process of extracting knowledge from data and discovering hidden relationships, which has recently been widely used in healthcare to predict various diseases [15–]. Some studies only used demographic risk factors (lifestyle and laboratory data) to predict breast cancer, while others relied on mammographic stereotypes or patient biopsy data [19–20]. Others demonstrated the use of genetic data in predicting breast cancer [21].

Developing a model that takes into account all known risk factors is a major challenge in predicting breast cancer [22]. Current prediction models may only analyze mammographic images or demographic risk factors, leaving out other important variables. Furthermore, these models, which are accurate enough to identify high-risk women, may require multiple screenings and invasive sampling using magnetic resonance imaging (MRI) and ultrasound. Patients may experience financial and psychological burdens [23].

Breast cancer risk can be effectively predicted using a variety of factors, including demographic, laboratory, and mammographic risk factors [24, 25]. As a result, multifactorial models that include a large number of risk factors can be more effective at assessing the risk of breast cancer [26]. The current study sought to predict breast cancer using various machine learning approaches while accounting for a variety of modelling variables.

4.3   METHODOLOGY

This study uses the Breast Cancer Dataset from kaggle.com, which is a well-known public data repository for machine learning and data science [27]. The dataset consisted of 569 entries and 32 columns or features. All the columns were used for training the machine learning (ML) model except the diagnosis variable, which was considered as output. The diagnosis variables consist of two values, 'M' for malignant and 'B' for benign. These values were mapped as 1 and 0 respectively for training the ML model.

The first step towards any classification model is the data preprocessing steps. This involves compensating for missing values, mapping all values with 'object' type into integers and finally standardizing all the entries to avoid any redundancy and ambiguity in data. This step ensures that the machine learning model will give accurate predictions.

The breast cancer dataset used in this study consists of no missing values. The dataset is split into training and testing sets with 30% data used for training the classification model. The entire dataset is standardized using the Standard Scaler method. The original dataset consists of 32 columns or features. However, all the features (such as id) may not be relevant in the diagnosis. Thus, the focus is on selecting the best features or variables that may significantly contribute to the cancer diagnosis. This study uses the RFE and RFECV feature selection methods to find the most significant features for cancer diagnosis. The recursive feature elimination (RFE) and RFE with cross-validation are used for this purpose.

4.3.1   STATISTICAL ANALYSIS

Statistical tests are important tools for analyzing data because they help to measure uncertainty, establish relationships, and use facts to make choices. Researchers can figure out the important differences or relationships among different variables of the dataset under consideration. Statistical tests help to analyze the differences between meaningful results and random variation. They offer a glimpse of how the different groups or conditions compare and which factors are related or important, and they help validate the choice of predictive model. This study used the IBM SPSS software to carry out the statistical analysis of breast cancer dataset.

Bayesian testing is performed on the dataset. In this test, the researchers instead of rejecting a null hypothesis, apply Bayes' theorem on the observed data to update the hypothesis. This method enables them to understand the data and the posterior probabilities for hypotheses instead of just p-values. When there is a lot of information to consider, a small sample size, or past knowledge that can help with the analysis, Bayesian testing is very helpful. Figure 4.2 shows the null hypothesis necessary to perform statistical tests on the data. Tables 4.1 and 4.2 show the results for the Bayesian test. Figure 4.3 shows the posterior distribution graph for the correlation coefficients.

The statistical inference regarding the proportion parameter is made by the one-sample binomial test by a comparison with a hypothesized value. Figures 4.4 to 4.12 show the histogram plots for some of the variables of the dataset. Figure 4.13 shows the results of one-sample binomial test performed on breast cancer data. The total sample size is 569 with test statistics of 212 and standard error of 11.927.

4.3.2 CLASSIFICATION MODELS

After data preprocessing and statistical analysis, the next steps involve splitting the data into training and testing samples using train_test_split. Training data is used to train the machine learning model to understand the patterns in the data and predict the results, the testing data is then used to check how accurate these predictions are. The first classification model used is the support vector machine (SVM) classifier. This classifier uses the 'rbf' kernel to make predictions. Then, we used the RFE-based SVM (RFE-SVM) model to predict the benign and malignant cases of breast cancer and lastly the RFE with cross validation based SVM (RFECV-SVM) model. Figure 4.14 shows the flow diagram for the entire methodology.

4.3.3 SUPPORT VECTOR MACHINES

SVM is one of the most widely used classification models for predictions. This model divides the dataset based on a hyperplane. This hyperplane is the one having maximum distance from both the classes. The points lying closest to hyperplanes are called support vectors. Now let us consider a random point 'X' and a vector 'w' which is perpendicular to this hyperplane. Let 'c' be the distance between the vector 'w' and the hyperplane. Then, we can decide if the point, X lies on the hyperplane, to the right or left of the hyperplane. Algorithm 1 summarizes these steps.

Algorithm 1 Support Vector Machine (SVM) Algorithm Start

Initialize SVM:

    Choose a kernel function (e.g., linear, polynomial, RBF)
    Set hyperparameters, such as C (regularization parameter) and kernel-specific parameters

Train SVM:

Use the training dataset to fit the SVM model
Update model parameters to minimize the classification error and maximize the margin
If $X \cdot \vec{w} = c$, then X
If $\vec{X} \cdot \vec{w} > c$, then $\vec{X}$ If $\vec{X} \cdot \vec{w} < c$, then $\vec{X}$

Predict:

Use the trained SVM model to make predictions on the test dataset
Assign labels based on the position of data points relative to the hyperplane

Output:

Output the predicted labels

Evaluation:

Assess the performance of the SVM classifier using evaluation metrics such as accuracy, precision, recall, F1-score, and/or confusion matrix

Interpretation:

Analyze evaluation metrics to understand the performance of the SVM classifier
Adjust hyperparameters or preprocessing steps as necessary based on evaluation results

Deployment:

Deploy the trained SVM model for classifying new, unseen samples **Stop**

4.3.4 RECURSIVE FEATURE ELIMINATION WITH CROSS-VALIDATION

RFECV is a feature selection methodology that exhibits notable methodological characteristics by integrating the precision of RFE with the adaptability of cross-validation. Initially, it employs a distinct RFE machine learning technique to meticulously prioritize the significance of each feature inside the dataset. The initial stage involves assessing the predictive impact of individual qualities, so repeatedly enhancing features by eliminating the least significant ones from the current set. This stage decreases the number of dimensions and avoids collinearity, leading to a subset of features that aligns with the intrinsic complexity of the dataset. The algorithm constructs a prognostic model following each exclusion, guaranteeing that the retained traits contribute to the model's overall efficacy. The algorithm converges iteratively to the optimal subset of features until it reaches the desired number of features.

Hypothesis Test Summary

	Null Hypothesis	Test	Sig.	Decision
1	The categories defined by diagnosis = 1 and 0 occur with probabilities 0.5 and 0.5.	One-Sample Binomial Test	.000[1]	Reject the null hypothesis.
2	The distribution of id is normal with mean 30,371,831 and standard deviation 125,020,585,612.	One-Sample Kolmogorov-Smirnov Test	.000[1]	Reject the null hypothesis.
3	The distribution of radius_1ean is normal with mean 14.127 and standard deviation 3.524.	One-Sample Kolmogorov-Smirnov Test	.000[1]	Reject the null hypothesis.
4	The distribution of texture_1ean is normal with mean 19.29 and standard deviation 4.301.	One-Sample Kolmogorov-Smirnov Test	.002[1]	Reject the null hypothesis.
5	The distribution of peri1eter_1ean is normal with mean 91.97 and standard deviation 24.299.	One-Sample Kolmogorov-Smirnov Test	.000[1]	Reject the null hypothesis.
6	The distribution of area_1ean is normal with mean 654.9 and standard deviation 351.914.	One-Sample Kolmogorov-Smirnov Test	.000[1]	Reject the null hypothesis.
7	The distribution of s1oothness_1ean is normal with mean 0.09636 and standard deviation 0.014.	One-Sample Kolmogorov-Smirnov Test	.076[1]	Retain the null hypothesis.
8	The distribution of co1pactness_1ean is normal with mean 0.10434 and standard deviation 0.053.	One-Sample Kolmogorov-Smirnov Test	.000[1]	Reject the null hypothesis.
9	The distribution of concavity_1ean is normal with mean 0.0887993 and standard deviation 0.080.	One-Sample Kolmogorov-Smirnov Test	.000[1]	Reject the null hypothesis.
10	The distribution of concavepoints_1ean is normal with mean 0.048919 and standard deviation 0.039.	One-Sample Kolmogorov-Smirnov Test	.000[1]	Reject the null hypothesis.
11	The distribution of sy11etry_1ean is normal with mean 0.1812 and standard deviation 0.027.	One-Sample Kolmogorov-Smirnov Test	.001[1]	Reject the null hypothesis.
12	The distribution of fractal_di1ension_1ean is normal with mean 0.06280 and standard deviation 0.007.	One-Sample Kolmogorov-Smirnov Test	.000[1]	Reject the null hypothesis.
13	The distribution of radius_se is normal with mean 0.4052 and standard deviation 0.277.	One-Sample Kolmogorov-Smirnov Test	.000[1]	Reject the null hypothesis.

Asymptotic significances are displayed. The significance level is .05.

[1]Lilliefors Corrected

FIGURE 4.2   Null hypothesis for breast cancer dataset.

4.4   RESULTS AND DISCUSSION

The dataset originally consisted of 569 samples and 32 features. The features selected after the SVM-RFECV are: concave points_mean, perimeter_se, radius_worst, texture_worst, area_worst, smoothness_worst and concave points_worst. Table 4.3 shows the classification metrics using all the three classification models. The SVM model is able to achieve an accuracy of 97.66%, while the RFE-SVM model achieves

TABLE 4.1
Bayesian Estimates of Coefficients

Parameter	Posterior			95% Credible Interval	
	Mode	Mean	Variance	Lower Bound	Upper Bound
diagnosis = 0	12.147	12.147	.016	11.896	12.397
diagnosis = 1	17.463	17.463	.028	17.138	17.788

TABLE 4.2
Bayesian Estimates of Error Variance

Parameter	Posterior			95% Credible Interval	
	Mode	Mean	Variance	Lower Bound	Upper Bound
Error variance	5.790	5.831	.121	5.189	6.551

Note: 0 is for Benign and 1 for malignant cases.

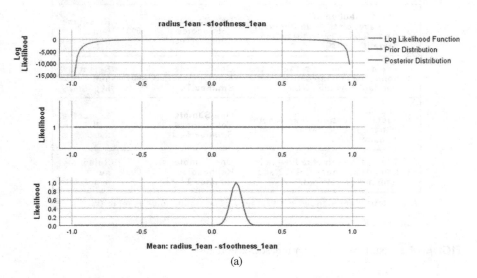

(a)

FIGURE 4.3   Posterior distribution graph for correlation coefficients.

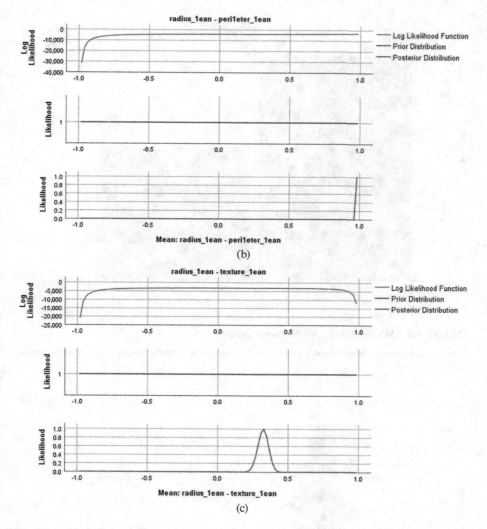

FIGURE 4.3 (*Continued*)

an accuracy of 97.36%. The best accuracy (98.25%) is achieved by the RFECV-SVM based classification model.

Figures 4.15 and 4.16 show the classification metrics, confusion matrix and ROC curve for all the three classification models. The SVM model was able to classify 103 and 64 cases of benign cancer and malignant cancer as true and only 2 cases in each category as false. The RFE-SVM model was able to correctly predict 144 and 78 cases of benign and malignant cases, but the number of false predictions is also high to 4 cases. RFECV showed the best results, with correct predictions of 105 benign and 63 malignant cases with 0 cases being falsely predicted. This proves its efficacy in the prediction of breast cancer diagnosis.

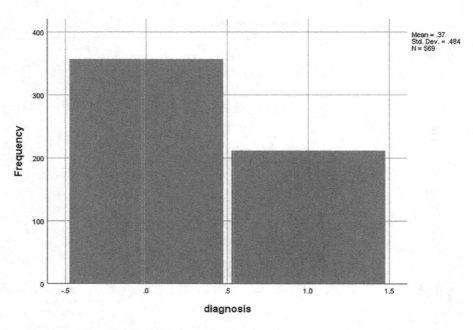

FIGURE 4.4   Histogram plot for diagnosis variable.

Note: The benign cases have a frequency of 350 and malignant cases have a frequency of 210 cases approx.

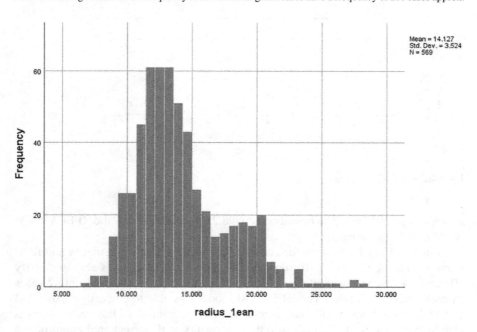

FIGURE 4.5   Histogram plot for radius of lobes variable.

Note: Values range between 6.98 and 28.1.

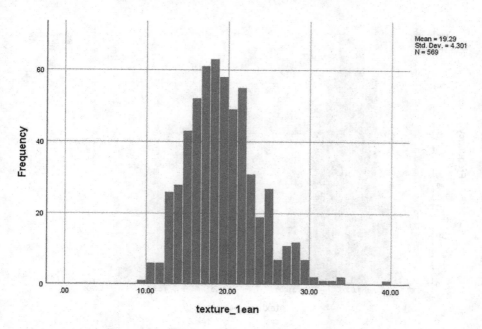

FIGURE 4.6   Histogram plot for mean of surface texture variable.

Note: Values range between 9.71 and 39.3.

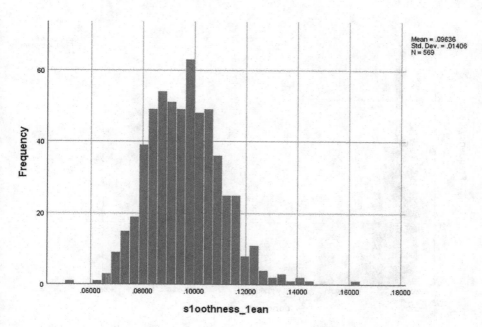

FIGURE 4.7   Histogram plot for mean of smoothness levels variable.

Note: Values range between 0.05 and 0.16.

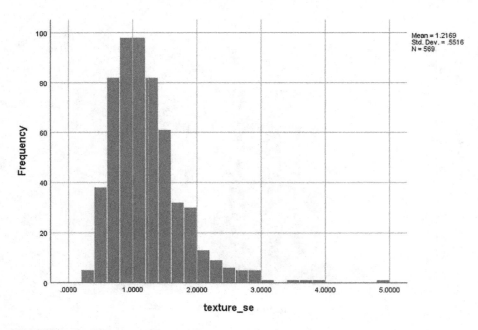

FIGURE 4.8   Histogram plot for SE of texture variable.

Note: Values range between 0.36 and 4.88.

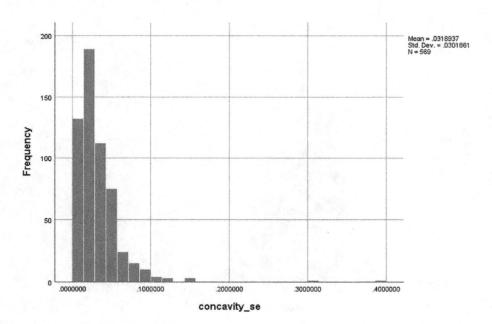

FIGURE 4.9   Histogram plot for SEE of concavity variable.

Note: Values range between 0 and 0.4.

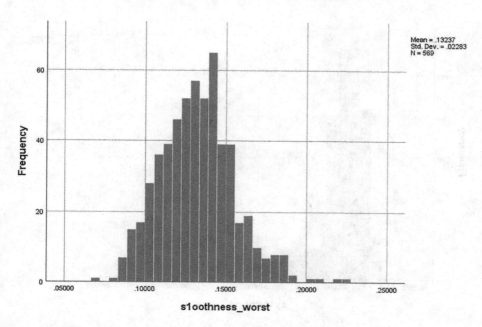

FIGURE 4.10 Histogram plot for worst smoothness variable.

Note: Values range between 0.07 and 0.22.

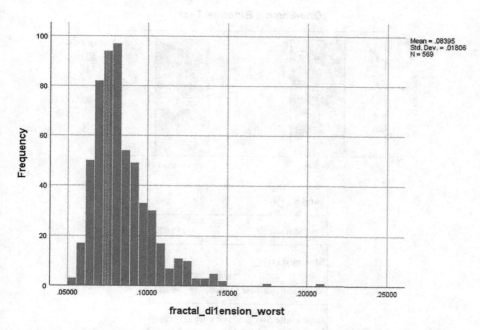

FIGURE 4.11 Histogram plot for Worst Fractal Dimension variable.

Note: Values range between 0.06 and 0.21.

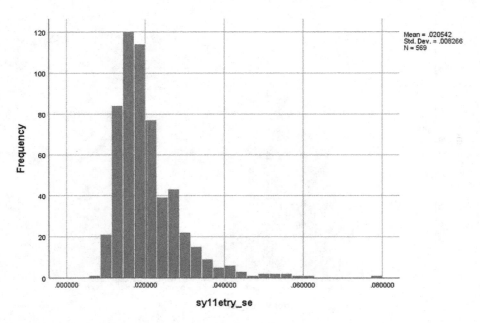

FIGURE 4.12   Histogram plot for SE of symmetry variable.

Note: Values range between 0.01 and 0.08.

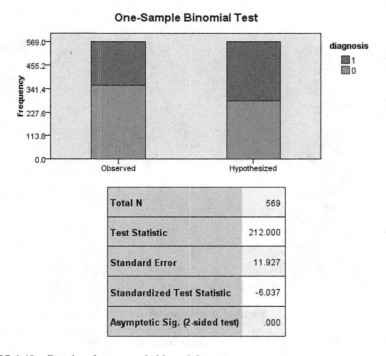

FIGURE 4.13   Results of one sample binomial tests.

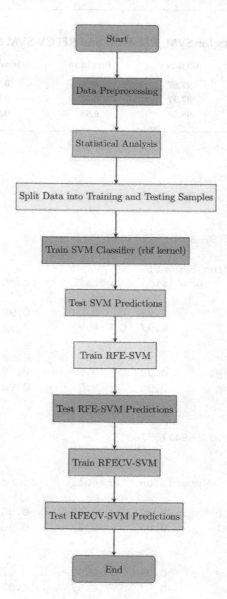

FIGURE 4.14 Flow diagram for entire methodology.

4.5 CONCLUSION

This chapter performed a comprehensive examination of exploratory data analysis using a breast cancer data set obtained from the renowned Kaggle dataset. The analysis begins by carrying out an in-depth investigation of the breast cancer dataset's attributes. This study discovered undetected trends, associations, and possible

TABLE 4.3
Classification Metrics for SVM, RFE-SVM and RFECV-SVM Models

Classification model	Accuracy	Precision	Sensitivity	F-score
SVM	97.67	0.98	0.97	0.98
RFE-SVM	97.37	0.96	0.97	0.95
RFECV-SVM	98.24	0.98	0.966	0.978

```
Confusion Matrix:
[[103    2]
 [  2   64]]

Accuracy: 0.9766081871345029

Classification Report:
              precision    recall  f1-score   support

           0       0.98      0.98      0.98       105
           1       0.97      0.97      0.97        66

    accuracy                           0.98       171
   macro avg       0.98      0.98      0.98       171
weighted avg       0.98      0.98      0.98       171
```
(a)

```
Accuracy: 0.9736842105263158
[[144    4]
 [  2   78]]
              precision    recall  f1-score   support

           0       0.99      0.97      0.98       148
           1       0.95      0.97      0.96        80

    accuracy                           0.97       228
   macro avg       0.97      0.97      0.97       228
weighted avg       0.97      0.97      0.97       228
```
(b)

FIGURE 4.15 Confusion matrices, accuracy and classification metrics using (a) SVM, (b) SVM-RFE and (c) SVM-RFECV classification algorithms respectively.

factors of influence from the collected information using data visualization, qualitative statistics, and relationship investigation between variables. Utilizing the method of statistical significance testing and different classification algorithms, the proposed methodology resulted in achieving optimal classification metrics. The SVM

```
Accuracy: 0.9824561403508771
[[105    0]
 [  3   63]]
                precision    recall  f1-score   support

           0         0.97      1.00      0.99       105
           1         1.00      0.95      0.98        66

    accuracy                            0.98       171
   macro avg         0.99      0.98      0.98       171
weighted avg         0.98      0.98      0.98       171
```

(c)

FIGURE 4.15 (Continued)

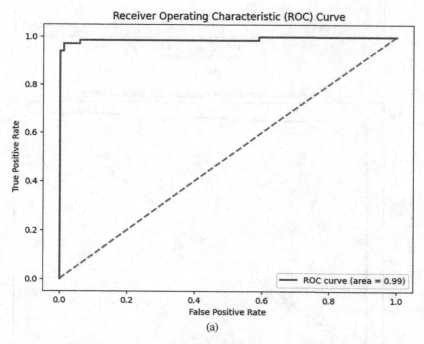

(a)

FIGURE 4.16 ROC curve using (a) SVM, (b) SVM-RFE and SVM-RFECV classification algorithms respectively.

algorithm alone was able to achieve an accuracy of 97.67% whereas after feature selection using RFE, the classification model achieved an accuracy of 97.37%. On the other hand, the RFE-based feature selection method after cross validation when combined with SVM achieved an accuracy of 98.24%. This proved the efficacy of

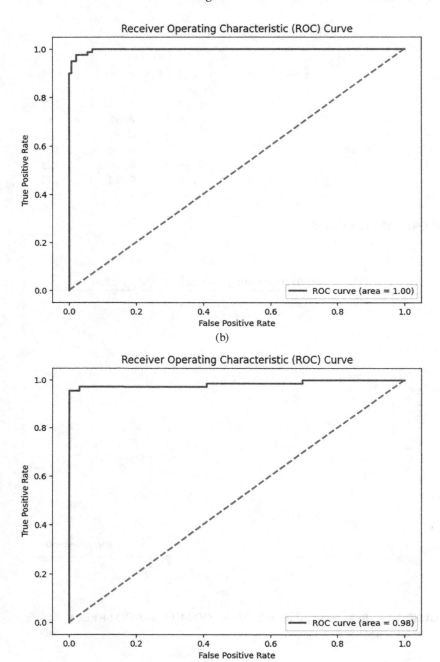

(b)

(c)

FIGURE 4.16   (*Continued*)

the RFECV-SVM model in accurately classifying gender-inclusive breast cancer into benign and malignant cases. Future work may involve exploring other complex machine learning algorithms that will also focus on reducing the prediction error.

REFERENCES

[1] Singh P, Kumar R, Gupta M, Al-Turjman F. SegEIR-Net: A robust histopathology image analysis framework for accurate breast cancer classification. Current Medical Imaging. 2024 Feb 9. doi: 10.2174/0115734056278974231211102917.

[2] Singh P, Kumar R, Gupta M, Obaid AJ. Transfer learning based breast cancer classification using histopathology images. In *2024 International Conference on Advancements in Smart, Secure and Intelligent Computing (ASSIC)* (pp. 1–4). IEEE, January 2024.

[3] El-Nabawy A, El-Bendary N, Belal NA. A feature-fusion framework of clinical, genomics, and histopathological data for METABRIC breast cancer subtype classification. Applied Soft Computing. 2020;91:20. doi: 10.1016/j.asoc.2020.106238.

[4] Nassif AB, Abu Talib M, Nasir Q, Afadar Y, Elgendy O. Breast cancer detection using artificial intelligence techniques: A systematic literature review. Artificial Intelligence in Medicine. 2022;127:13. doi: 10.1016/j.artmed.2022.102276.

[5] Yao H, Zhang X, Zhou X, Liu S. Parallel structure deep neural network using CNN and RNN with an attention mechanism for breast cancer histology image classification. Cancers. 2019;11:1901. doi: 10.3390/cancers11121901.

[6] Ha R, Chang P, Mutasa S, Karcich J, Goodman S, Blum E, Kalinsky K, Ms MZL, Jambawalikar S. Convolutional neural network using a breast MRI tumor dataset can predict Oncotype Dx recurrence score. Journal of Magnetic Resonance Imaging. 2019;49:518–524. doi: 10.1002/jmri.26244.

[7] Bhatt C, Kumar I, Vijayakumar V, Singh KU, Kumar A. The state of the art of deep learning models in medical science and their challenges. Multimedia System. 2021;27:599–613. doi: 10.1007/s00530-020-00694-1.

[8] Fujioka T, Mori M, Kubota K, Oyama J, Yamaga E, Yashima Y, Katsuta L, Nomura K, Nara M, Oda G, et al. The utility of deep learning in breast ultrasonic imaging: A review. Diagnostics. 2020;10:1055. doi: 10.3390/diagnostics10121055.

[9] Aavula R, Bhramaramba R, Ramula US. A comprehensive study on data mining techniques used in bioinformatics for breast cancer prognosis. Journal of Innovation in Computer Science and Engineering. 2019;9(1):34–39.

[10] Chaurasia V, Pal S, Tiwari BB. Prediction of benign and malignant breast cancer using data mining techniques. Journal of Algorithms & Computational Technology. 2018;12(2):119–126. doi: 10.1177/1748301818756225.

[11] Fan J, Wu Y, Yuan M, Page D, Liu J, Ong IM, Peissig P, Burnside E. Structure-leveraged methods in breast cancer risk prediction. The Journal of Machine Learning Research. 2016;17(1):2956–2970.

[12] Stephens K. New mammogram measures of breast cancer risk could revolutionize screening. AXIS Imaging News. 2020.

[13] Guan Y, Nehl E, Pencea I, Condit CM, Escoffery C, Bellcross CA, McBride CM. Willingness to decrease mammogram frequency among women at low risk for hereditary breast cancer. Scientific Reports. 2019;9(1):9599. doi: 10.1038/s41598-019-45967-6.

[14] Jothi N, Husain W. Data mining in healthcare—a review. Procedia Computer Science. 2015;72:306–313. doi: 10.1016/j.procs.2015.12.145.

[15] Maxwell K, Nathanson K. Common breast cancer risk variants in the post-COGS era: A comprehensive review. Breast Cancer Research. 2013;15(6):212. doi: 10.1186/bcr3591.

[16] McCarthy AM, Keller B, Kontos D, Boghossian L, McGuire E, Bristol M, et al. The use of the Gail model, body mass index and SNPs to predict breast cancer among women with abnormal (BI-RADS 4) mammograms. Breast Cancer Research. 2015;17(1). doi: 10.1186/s13058-014-0509-4.

[17] Kumar R, Gupta M, Yadav A, Gautam M. Breast cancer prediction using deep learning models. In 2023 International Conference on Circuit Power and Computing Technologies (ICCPCT) (pp. 1275–1279). Kollam, India, 2023. doi: 10.1109/ICCPCT58313.2023.10245816.

[18] Williams K, Idowu PA, Balogun JA, Oluwaranti AI. Breast cancer risk prediction using data mining classification techniques. Transactions on Networks and Communications. 2015;3(2):1–11. doi: 10.14738/tnc.32.662.

[19] Oyewola D, Hakimi D, Adeboye K, Shehu MD. Using five machine learning for breast cancer biopsy predictions based on mammographic diagnosis. International Journal of Engineering Technologies. 2016;2(4):142–145. doi: 10.19072/ijet.280563.24.

[20] Hajiloo M, Damavandi B, Hooshsadat M, Sangi F, et al. Breast cancer prediction using genome wide single nucleotide polymorphism data. BMC Bioinformatics. 2013;14(Suppl 13):S3. doi: 10.1186/1471-2105-14-S13-S3.

[21] Hou C, Zhong X, He P, Xu B, Diao S, Yi F, Zheng H, Li J. Predicting breast cancer in Chinese women using machine learning techniques: Algorithm development. JMIR Medical Informatics. 2020;8(6):e17364. doi: 10.2196/17364.

[22] Koopmann BDM, Harinck F, Kroep S, Konings ICAW, Naber SK, et al. Identifying key factors for the effectiveness of pancreatic cancer screening: A model-based analysis. International Journal of Cancer. 2021;149(2):337–346. doi: 10.1002/ijc.33540.

[23] Arefan D, Mohamed AA, Berg WA, Zuley ML, Sumkin JH, Wu S. Deep learning modeling using normal mammograms for predicting breast cancer risk. Medical Physics. 2020;47(1):110–118. doi: 10.1002/mp.13886.

[24] Juneja A, Kumar R, Gupta M. Smart healthcare ecosystems backed by IoT and connected biomedical technologies. In 2022 Fifth International Conference on Computational Intelligence and Communication Technologies (CCICT) (pp. 230–235). IEEE, July 2022.

[25] Behravan H, Hartikainen JM, Tengström M, Kosma VM, Mannermaa A. Predicting breast cancer risk using interacting genetic and demographic factors and machine learning. Scientific Reports. 2020;10(1):11044. doi: 10.1038/s41598-020-66907-9.

[26] Siegel RL, Miller KD, Fuchs HE, Jemal A. Cancer statistics, 2022. CA Cancer Journal for Clinicians. 2022;72(1):7–33. doi: 10.3322/caac.21708.

[27] Kaggle. Breast cancer dataset. Kaggle, n.d. www.kaggle.com/datasets/yasserh/breast-cancer-dataset.

5 Unlocking Gender-Based Health Insights with Predictive Analytics

*Vinod Kumar, Chander Prabha,
and Md. Mehedi Hassan*

5.1 BACKGROUND

This chapter discusses the potential of AI and predictive analytics to uncover gender-based health insights in the quickly changing sector of healthcare. The practices and the future of healthcare will be greatly influenced by gender-specific health data. The authors emphasize the need to recognize and correct errors, such as improper selection and observation bias when using large data sets. They exhort readers to view bias with objectivity and to classify all effects as either beneficial or detrimental [1]. Encouraging public participation can significantly reduce knowledge gaps and promote inclusiveness in the field of health. This chapter covers several countries, including the United States, Canada, Pakistan, Australia, and India [2]. The authors navigate the challenging worlds of artificial intelligence, machine learning, and big data in healthcare to promote a complete knowledge of health that challenges conventional assumptions.

One socially imposed characteristic that determines roles, norms, and relationships between men and women in a society is gender. These concepts are culturally specific and subject to change. Discrimination and social rejection can result from breaking gender norms, and these things can be harmful to an individual's health. The power imbalances that frequently result from the hierarchical structure of genders also affect other social and financial dimensions of inequality. The combination of gender curation, machine learning, and artificial intelligence has been the focal point of the paradigm change in healthcare in recent years. The expansion of gender-based healthcare appears to be primarily driven by the transformative possibilities of precision medicine and predictive analytics [3]. This chapter explores the significance of computer health research in identifying gender disparities that go beyond social and biological factors, as observational retrospective studies demonstrate the transformative power of technology-induced inequality. Large data sets play a significant role in health because they can shed light on the mechanisms underlying both disease and health. However, to examine the implicit biases found in large data sets, it is essential to have an impartial viewpoint, which may include both observational and decision biases. An AI-based system that forecasts patient outcomes by examining the medical data of a particular patient Radha Devi is shown in Figure 5.1.

DOI: 10.1201/9781003473435-5

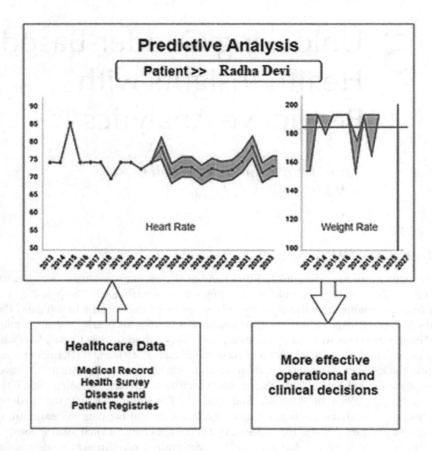

FIGURE 5.1   AI predictive analysis.

This contains information from her health survey about the diagnosed illness as well as her consistent enrolment in the hospital administration system. This system's integration seems to be greatly beneficial in guiding clinical and operational choices. Additionally, depending on their level of influence, it encourages proactive measures aimed at minimizing undesirable biases and maximizing preferred biases, which improve diagnostic accuracy. Social media communication has the power to eliminate knowledge gaps and encourage inclusiveness [4]. It seeks to facilitate inclusive practices, a paradigm shift in research techniques, and the development of a holistic vision of health that goes beyond traditional expectations by navigating the complicated world of artificial intelligence, machine learning, and enormous amounts of data in healthcare. Predictive analytics and cutting-edge AI methods need to be applied extensively to create a healthcare system that is more inclusive and enhances the care given to women.

Wearable technology uses machine learning algorithms for time series analysis, allowing for the tracking of health trends and activity detection. Conversely, the International Medical Journal relies on Bayesian networks and support vector machines (SVM) for online patient consultation and monitoring. Eventually, techniques like sentiment analysis and machine learning for sentiment recognition can identify and monitor feelings in the assessment of mental health.

5.2   THE SOCIO-ECOLOGICAL GENDER-BASED HEALTHCARE MODEL

Using machine learning techniques to improve the global health of all sexes has opened up new possibilities, especially in terms of tackling gender-based health disparities. A worldwide framework for comprehending health outcomes is provided, one that considers the social, interpersonal, communal, and individual elements that influence health and well-being. This is made possible with the socio-ecological model. The author examines here the interplay of machine learning, algorithmic justice, and the socio-ecological model in the context of gender-based healthcare.

5.2.1   Gender-Based Health Disparities in Machine Learning

Addressing health disparities, particularly those rooted in gender, race/ethnicity, and income levels, has become a central focus in public and population health research. Racial and ethnic health disparities refer to variations in health status and the distribution of health resources among different population groups. These differences stem from the social conditions in which individuals are born, grow, live, work, and age. The impact of these disparities is often magnified within specific gender, income, and race/ethnicity groups, leading to unequal health outcomes and risks [5].

In recent times, the intersection of machine learning and healthcare has brought forth the crucial consideration of algorithmic fairness. Algorithmic fairness is a burgeoning field within machine learning, aiming to address and mitigate differences in outcomes across various social groups. Figure 5.2 highlights the many levels that affect health outcomes.

At the macro level, the foundation of health systems is public policy, which includes international, national, municipal, and government laws and regulations. The community level includes the evolving relationships among various businesses and groups that impact health access and distribution. The structures and procedures that affect how healthcare is provided and accessed inside businesses and social institutions are identified at the organizational level. Upon approaching the individual, the interpersonal level considers the influence of friends, family, and social networks on health-related decisions and behaviors. Ultimately, at a micro level, an individual's knowledge, attitudes, and skills have an influence. This includes ensuring fairness in machine learning predictions and recommendations related to health.

5.3   ALGORITHMIC APPROACHES

Statistical Notions: Algorithmic approaches employ statistical notions to guarantee some form of parity for members of different protected groups, such as race

FIGURE 5.2   Levels that affect health outcomes in socio-ecological gender-based healthcare.

and gender. These are used to eliminate biases. Also ensures that the findings are correct.

- **The Interpersonal Level:** These are improved ML algorithms. By taking into consideration the language and genre-related information, the natural language processing (NLP) algorithms assist in the analysis of patients. Resulting in the contribution of the development of a transparent health environment by taking into account the interpersonal factors affecting health research activities [7].
- **Individual Notions:** Focusing on ensuring that particular individuals (deemed similar in the context) w.r.t specific classification task get reasonable results. The aim is to prevent discrimination against particular individuals having similar health characteristics, irrespective of their demographic factor of individual or gender [6].

- **Identifying Economic and Social Factors:** Communities are greatly affecting the state of health. The ability of ML models to analyze massive data helps in identifying socio-economic factors that influence gender-based health abnormalities. Public health initiatives can be deployed to tailor the address gaps at source and ensure access to healthcare [8].
- **Social Considerations:** At the societal level, algorithmic approaches are becoming more significant. The gender-based health abnormalities could be maintained by bias in health algorithms [9].

5.4 AI-DL MODEL FOR GENDER-BASED HEALTHCARE CLASSIFICATION

A sophisticated DL model utilizing a hybrid architecture is proposed that enhances the genre-based health categorization by combining the benefits of GoogLeNet, AlexNet, and ResNet. The aim is to leverage the distinctive features and capabilities of each network architecture to enhance the overall classification's performance.

5.4.1 FUSION MODEL OF ALEXNET, GOOGLENET, AND RESNET

- AlexNet is taken for picture classification and feature extraction tasks due to several convolution layers. AlexNet ensures that the model captures essential intricate details for sex-based health classification [10].
- GoogLeNet (Inception): This uses initiation modules, thus enabling the network to find features at various scales. Its integration into the model improves its capacity to recognize intricate structures and patterns in medical pictures [11].
- The ResNet makes the gradient disappear for creating extraordinarily deep networks. Its integration makes it feasible to find out hierarchical characteristics efficiently [12].

Figure 5.3 gives an insight into the process of classifying healthcare patients (gender-based) using an AI-DL model. It presents a novel hybrid approach incorporating AlexNet, GoogLeNet, and ResNet. In the first phase, every component's features are extracted. These extracted features are then combined with DenseNet (DL framework). The unique features are then combined that are gathered by each model. Ultimately, the categorization results serve as input from this integrated model, offering a thorough analysis based on genre.

The extracted features from AlexNet, GoogLeNet, and ResNet are merged and concatenated with DenseNet. It does this by enhancing the model's ability to figure out complex dependencies in the data [13] [14].

5.4.2 GENDER-BASED HEALTH CLASSIFICATIONS

The SoftMax activation is used on these combined features for sex-based health categorization by transferring these into a classification layer. A labeled dataset is used for training. The model picks up and applies a variety of patterns and features taken

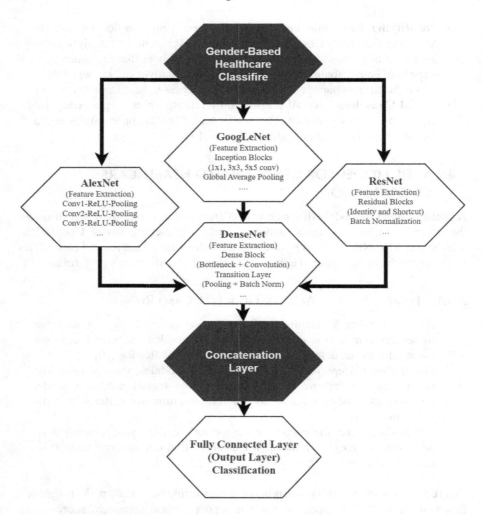

FIGURE 5.3   Gender-based healthcare classification using AI-DL model.

from medical images [15]. Table 5.1 provides a deep review of gender-specific health problems together with the equivalent AI investigating methods.

To improve the predictive accuracy and generalizability to a wider range of health issues, this all-encompassing hybrid deep learning model makes use of the strengths of different architectures to capture a variety of variables relevant to gender-based disorders. For prostate cancer, the focus is on analyzing medical images for unique characteristics related to the prostate region, using ResNet to identify abnormalities in the prostate tissue. The same is true for ovarian cancer, where a specific investigation of irregularities in the ovary region is carried out, and GoogLeNet is used to find subtleties of ovarian tissue. Utilizing AlexNet to evaluate breast regions for certain features, patterns in breast tissue are emphasized and lumps or malignancies

TABLE 5.1
Gender-Based Healthcare Challenges and AI Approaches

Gender-Based Challenges	AlexNet	GoogLeNet	ResNet	Dense Net
Prostate Cancer (PC) [16]	Examining medical images for specific characteristics related to the prostate area.	Multi-scale feature extraction using inception blocks.	Using residual connections to identify anomalies in prostate tissue.	Getting dense features out of the prostate area.
Ovarian Cancer (OC) [17]	Specialized study of anomalies in the ovary area of ovarian cancer.	Recognizing subtleties associated with ovarian tissue.	Using residual blocks to identify anomalies in ovarian structure.	Extracting dense characteristics particular to ovarian anomalies.
Breast Cancer (BC) [18]	Highlights patterns in breast tissue and detects lumps or cancers.	Employing inception blocks to analyze breast areas to obtain detailed characteristics.	Remaining connections are used to identify irregularities in breast tissue.	Getting dense features out of breast tissue.
Lung Cancer (LC) [19]	Analysis of irregularities in lung structure, with a focus on nodule identification.	A comprehensive examination of lung areas with different convolutional layers.	Use of residual connections to identify anomalies in lung tissue.	Extracting dense characteristics particular to anomalies of the lungs.
Nutritional Deficiencies (ND) [20]	Examining issues related to the skin, hair, and nails that may indicate shortages.	Identifying subtle characteristics linked to outward manifestations of malnutrition.	Using residual connections, anomalies linked to nutritional deficits are detected.	Identifying indicators of weaknesses by extracting dense features from images.

are found in cases of breast cancer. Regarding prostate cancer, the analysis focuses on abnormalities in the prostate's structure, and ResNet uses residual connections to find anomalies in the prostate tissue. Finally, under the framework of nutritional deficiencies (ND), issues related to skin, hair, and nails that indicate deficiencies are examined using DenseNet, which extracts dense features from pictures to identify deficiencies.

5.5 GENDER-BASED HEALTHCARE PREDICTIVE ANALYTICS CHALLENGES

With an emphasis on malnutrition specifically, the authors cover a wide variety of gender-based health-related issues in India, Canada, Pakistan, Australia, United State of America. Vitamin B12 insufficiency is recognized as a global problem that affects people in every area. Deficits in vitamin A still pose a threat, particularly to kids enrolled in maternal schools. Nonetheless, newborns are in danger from Iodine

TABLE 5.2
Medical Imaging Findings across Modalities

Gender-based ailments	Artificial Intelligence Influences Characteristics
PC [16]	Abnormalities in the prostate shape or size of MRI or ultrasound images
OC [17]	Identification of masses or cysts in pelvic ultrasound images
BC [18]	Detection of microcalcifications or masses in mammograms
LC [19]	Identification of lung nodules or lesions in CT scans
ND [20]	Analysis of skin, hair, and nails for signs of deficiencies

deficits, which highlights the need of iodizing salt everywhere [21]. An indication of how complicated nutritional problems are in India is the prevalence of folic acid and vitamin D deficiency in expecting mothers. These gaps must be closed, and to enhance overall health outcomes, comprehensive initiatives involving food preservation and health education are required. The authors have provided details to describe the entire gender-based description. Table 5.2 details attributes associated with each gender-based ailment. The authors also suggest possible areas of improvement.

Along with offering insights into the exact features or patterns in medical images that the deep learning model can identify for each health issue. The features described are unique to each gender-specific illness and could account for various methods of diagnosis or imaging technologies. The objective of these characteristics is to identify patterns or anomalies in medical imaging that suggest the existence of specific medical disorders. During the training phase, the deep learning model may make use of these attributes to improve its capacity to identify and categorize a variety of patterns linked to gender-based illnesses.

5.5.1 BREAST CANCER

Breast cancer (BC) is the most common kind of cancer among women globally; it will overtake lung cancer as the primary cause of cancer incidence worldwide in 2020. Roughly 2.3 million new cases were recorded in that year alone, accounting for 11.7% of all instances of cancer. Epidemiological studies estimate that by 2030, the worldwide burden of BC will be close to 2 million cases [22].

In India, the number of cases of breast cancer has increased significantly, increasing by about 50% between 1965 and 1985. Of the approximately 118,000 new cases that were recorded in 2016, 98.1% involved females. 526,000 instances were reported as the most common cases in the same year [23]. The age-standardized incidence rate of BC in females increased by 39.1% between 1990 and 2016, which is a significant rise over the previous 26 years. Every state in the union had a constant increase in this regard.

According to GLOBOCAN statistics for 2020, BC accounted for 10.6% (90,408) of all fatalities and 13.5% (178,361) of all cancer cases in India. In India, the total chance of developing breast cancer was estimated to be 2.81. The data presented here highlights the noteworthy influence of breast cancer on the nation's incidence and death rates. Understanding the prevalence and patterns of breast cancer becomes essential for developing effective healthcare interventions and policies as they dive

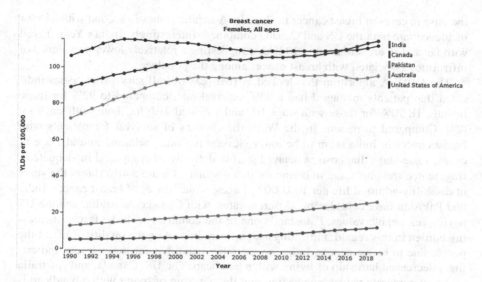

FIGURE 5.4(A)   Years lived with disability per 100,000 (all ages combined) female breast cancer cases.

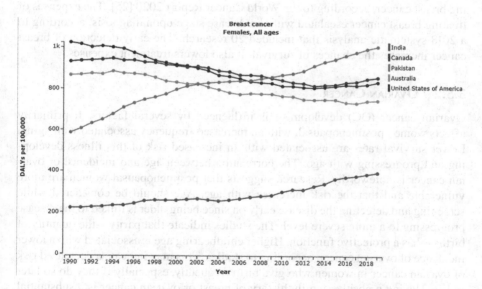

FIGURE 5.4(B)   Years of disability-adjusted life per 100,000 (all ages females) breast cancer cases.

further into the larger panorama of gender-based health insights. The data on Years Lived with Disability per 100,000 (all ages combined) Female Breast Cancer Cases is shown in Figure 5.4(a). In contrast to the United States, where a small decline in occurrences of breast cancer has been seen, the trend analysis shows a notable increase in cases of breast cancer in Canada. Australia comes in second with a lower incidence rate than the US and Canada combined. Interestingly, India's Years Lived with Disability are still around 10 lakhs, suggesting a relatively lower prevalence of infirmities associated with breast cancer among the populace.

The GBDLA algorithm that looked at five-year overall survival statistics indicated that patients in stage I had a 95% survival rate, compared to 92% for those in stage II, 70% for those with stage III, and a dismal 21% for those with stage IV [24]. Compared to nations in the West, the chances of survival for patients with breast cancer in India seem to be lower. Several reasons, including an older age of onset, a late-stage diagnosis, a delayed start to definitive therapy, and inadequate or fragmented treatment, are to blame for this disparity. Figure 5.4(b) shows the years of disability-adjusted life per 100,000 (all ages) female cases of breast cancer. India and Pakistan have considerably different rates, with Canada, Australia, and the US having reasonable values. Pakistan is one of the countries with the fastest increasing burden in this regard, indicating a significant impact on the quality of life of the people due to breast cancer. India, on the other hand, has far lower levels, indicating a decreased hardship of living with a handicap. The US, Canada, and Australia all show moderate values, suggesting that the hardship of living with a handicap is spread evenly among these three countries.

Early identification and fast treatment are the most successful strategies for reducing breast cancer, according to the World Cancer Report 2020 [25]. The expenses of treating breast cancer escalated with the disease stage upon diagnosis, according to a 2018 systematic analysis that included 20 research. The early detection of breast cancer increases the chances of survival. It also lowers treatment expenses.

5.5.2 Ovarian Cancer

Ovarian cancer (OC) development is influenced by several factors. It primarily affects women postmenopausal, with an increased frequency associated with aging. Lower survival rates are associated with an increased risk of this illness developing and progressing with age. The correlation between age and incidence of ovarian cancer is noteworthy. Research suggests that postmenopausal women are more vulnerable and that the risk increases with age. Age should be considered while screening and detecting the disease early on since being older is linked to the disease progressing to a more severe level. The studies indicate that parity—the quantity of births—has a protective function. Higher childbearing age is associated with a lower incidence of ovarian cancer, according to studies. This may indicate a decreased risk of ovarian cancer in women who give birth frequently, especially if they do so later in life. Having a positive family history of breast or ovarian cancer is a substantial and well-established risk factor. One's chance of acquiring ovarian cancer rises if they have close relatives who have had various types of cancer. It is also important to know one's family history of cancer because a personal history of the disease increases the risk. The association between smoking and ovarian cancer has been investigated in a few research articles, and the results indicate that smoking is associated with a higher risk, especially for tumours [26]. To lower the risk of subtypes of ovarian cancer, it is crucial to address lifestyle variables, such as quitting smoking.

With ovarian cancer accounting for 1.2% of all cancer cases in 2020, it is a serious health problem. With over 21,750 new cases recorded, cancer is still a very difficult disease to treat. Regretfully, the cost is reflected in the 13,940 fatalities linked to ovarian cancer that are projected to have occurred. Its etiology must be understood to develop efficient prevention measures and early detection techniques. Ovarian cancer has a five-year proportionate survival rate of 48.6%, which emphasizes the need for early detection. Because 58% of patients are discovered at a stage where metastases have already formed, this worrying figure causes the 5-year survival rate to plummet to 30.2%. However, once it is found locally, the survival rate soars to 92.6% [27]. This underlines how important early detection of ovarian cancer is.

Despite novel treatment techniques and current clinical trials, ovarian cancer is still a difficult illness that presents a substantial threat to women's health. Lack of efficient early detection measures results in delayed diagnosis and restricted treatment choices, which is the main barrier causing poor clinical outcomes. Furthermore, patient outcomes may be negatively impacted by clinical practice deviations from approved norms, underscoring the need for better ovarian cancer care strategies.

The creation of successful tactics is required in the quest for earlier and more treatable ovarian cancer diagnosis. After surgery, the amount of remaining disease is a critical determinant that affects patient survival. To maximize results, this complex treatment should only be carried out by highly skilled gynaecologist working in busy, high-volume environments, including big hospitals that see over 20 patients a year [28]. The evidence presented in Figure 5.5(a) indicates regional variations in the YLD rates for instances of ovarian cancer. The United States and Canada show a sharp rise in YLD rates, especially when compared to Canada, where a little decline is noted. Australia and Pakistan show intermediate degrees of existence between these extremes, in contrast, India shows a major growing tendency akin to that of the USA.

Handling ovarian cancer critically depends on patient empowerment via shared decision-making. Patients gain from a thorough conversation encompassing possible advantages, safety profiles, symptom management, and prognosis when contemplating recently developed therapy techniques or enrolling in clinical trials. Optimal care of patients requires a cooperative strategy between medical oncologists, therapeutic oncologists, and an interdisciplinary healthcare group.

Patients with ovarian cancer benefit greatly from early adoption of supportive services, which helps to optimize their course of therapy. Palliative medicine supports an additional, individualized, and holistic strategy for therapy by managing symptoms, offering support, and improving the overall state of life. In contrast, Figure 5.5(b) depicts the DALYs per 100,000 female ovarian cancer patients in the same locations, offering an outcome that is slightly different than Figure 5.5(a). Although there are some similarities between the patterns, the DALYs metric provides an additional perspective on the burden of ovarian cancer by accounting for years lived with a disability as well as years lost to early death.

Adequate and reasonably priced follow-up measures are necessary for patients in therapeutic remission. It is essential to educate patients about potential indicators of illness recurrence in addition to normal surveillance. In addition, promoting genetic risk counselling for patients, especially if it hasn't been done before in the early phases of the illness, can support an additional proactive effort and individualized approach to treatment.

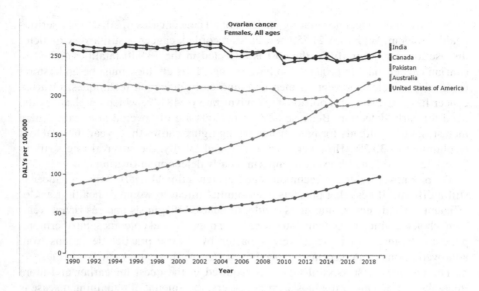

FIGURE 5.5(A)   Regional variations in the YLD rates for instances of ovarian cancer.

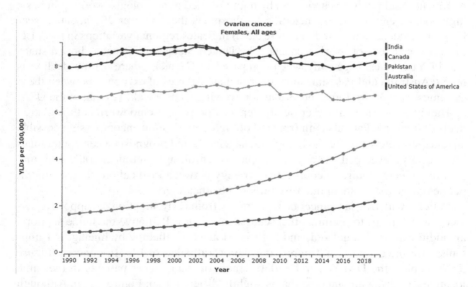

FIGURE 5.5(B)   Years of disability-adjusted life per 100,000 (all ages females) ovarian cancer cases.

5.5.3 Prostate Cancer

The use of artificial intelligence in prostate cancer treatment remains in its infancy. Early findings from AI models are promising for several elements of prostate cancer (PC) treatment. The development of sophisticated AI models that will be important for enhanced training, superior surgical results, diagnosis, and the creation of pre-dictive tools [40] for prostate cancer will require continuing training and testing of AI algorithms. Although there are many potential advantages to using AI in the treatment of prostate cancer, it is crucial to stress how crucial it is to preserve gov-ernance in the field of AI-related care [29]. To maintain morale and security proce-dures in the treatment of prostate cancer, prudent deployment of AI technology is essential. Figure 5.6(a) shows over the years lived with disability (YLD) per 100,000 males with prostate cancer. The statistics demonstrate striking differences across the nations; Pakistan exhibits a notable increase in comparison to Canada and the USA, while India and Australia are displaying remarkably high rates. Figure 5.6(b) displays the number of years of disability-adjusted life (DALYs) for every 100,000 male individuals diagnosed with prostate cancer. The results demonstrate that there are regional differences in the distribution, with Pakistan and the USA seeing rather comparable increases. While Canada and Australia also record incidences of gender-based differences in prostate cancer incidence, India reports modest rates, ranging from 60k to 80k.

Men are more likely than women to receive a prostate cancer diagnosis, and the number of cases each year is rising. The financial stability and resources in medical institutions that cure prostate cancer have been strained by the increase in cases as well as by the introduction of novel treatment choices and improved diagnostic tech-niques. Artificial Intelligence (AI) integration is a promising complement to human engagement and in certain cases, a possible replacement [39]. This may lessen some of the problems associated with resources, which would save costs [30]. Moreover, auto-mation tackles problems like as differences in observations across time and between various people. It is also capable of quickly and precisely analyzing large datasets. Examine the growing uses of AI and its related domains concerning prostate cancer.

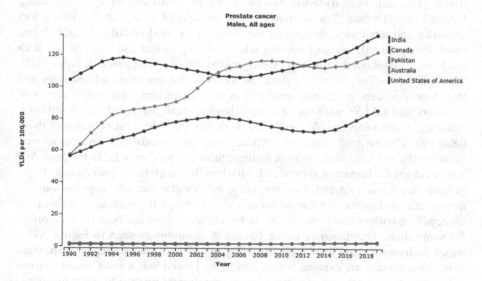

FIGURE 5.6(A) Years lived with disability.

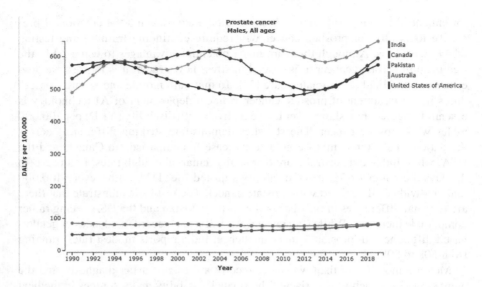

FIGURE 5.6(B) Years of disability-adjusted life per 100,000 (all ages males) prostate cancer cases.

5.5.4 LUNG CANCER

The fascinating history of lung cancer (LC) has spread around the world and had a significant impact on people's health and well-being. The research delves into the complexities of the LIDC-IDRI database to solve the puzzles surrounding this illness. The authors analyze the disorderly course of this horrifying illness using DICOM imaging data. The reference region consists of 21 regions, including 195 countries and territories, throughout the diagnostic period of 2015 to 2023. It has carefully examined data on longevity related to lung cancer and years of life with disability throughout this time to investigate the effects of various risk factors [31]. Through the darkness of this widespread sickness, the age-standardized rates and their yearly percentage changes emerged as beacons of hope. The major causes of mortality and DALYs were found to be behaviors, both ingrained and modifiable. Smoking, second-hand smoke, and a low-fruit diet were the most common of these behaviors. The harmful effects of environmental and occupational hazards as well as metabolic risks, particularly high fasting glucose values, were far in the past. Age has increased the impact of various risk variables. In the global impact mosaic, high-income North America and Asia are where behavioural hazards have the longest-lasting effects. Governments must act quickly to change the trajectory of this risk, especially in nations like Pakistan and India where smoking has been more common for some time. The disparity in the burden of disability is seen in Figure 5.7(a), which focuses on years lived with Disability per 100,000 cases. Gender disparities in health outcomes are evident, with males in the United States and Canada bearing a disproportionate amount of the burden of impairment relative to other locations.

On the other hand, India has the fewest disability-adjusted life years (DALYs) per 100,000 instances, which suggests that its health results are more favorable. Falling in the middle, Pakistan and Australia demonstrate regional variations in gender-based healthcare while also indicating a moderate burden of impairment. However, Figure 5.7(b) illustrates the years of life adjusted for the illness by 100,000 instances of women, emphasizing the impact of the illness and premature death. This time, the pattern is the same, but there are some slight variations. Though the number of years that women in the United States and Canada have adapted to disability is more than 100,000 instances, this difference is less pronounced when compared to males. India is the country with the lowest burden of years lived with a handicap adjusted for life, which suggests that women with ovarian cancer would have better health outcomes. Pakistan and Australia both retain a moderate level of disability-adjusted life years, which highlights the necessity of ongoing efforts to eliminate regional inequities in healthcare based on gender.

With an emphasis on men, Figure 5.7(c) illustrates the differences in Years Lived with Disability per 100,000 cases. There are gender-based health disparities in the United States and Canada, where men experience a greater amount of impairment in comparison to other locations. In contrast, India has the lowest Years Lived with Disability per 100,000 instances, indicating that its health results are generally good. Australia and Pakistan are in the middle, showing regional differences in healthcare for women and showing a modest burden of impairment. The burden of disease and early mortality, however, is shown by Figure 5.7(d), which shows the Years Lived with Disability per 100,000 instances for women. Though with a few minor differences, similar patterns appear. Though the difference is not as great as it is for men, women in the US and Canada live with a handicap for a significant number of years

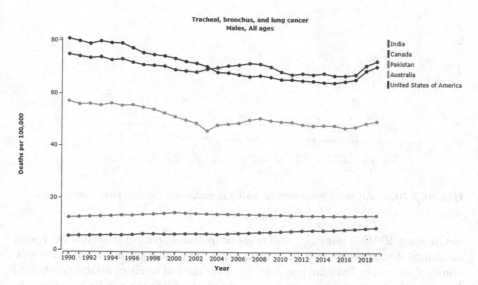

FIGURE 5.7(A)   All cases of death in males per 100,000 lung cancer cases.

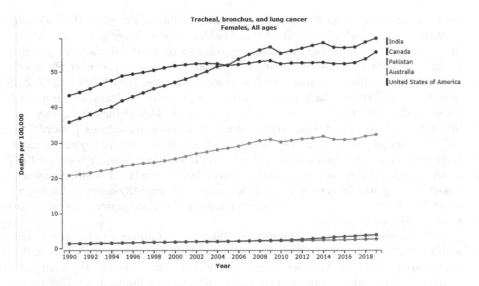

FIGURE 5.7(B)   All cases of death in females per 100,000 lung cancer cases.

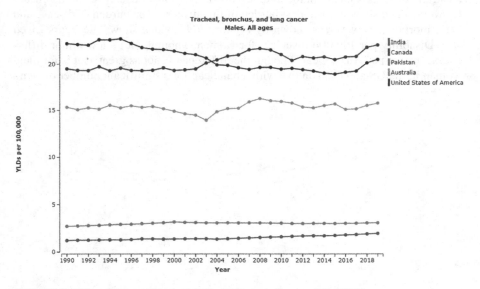

FIGURE 5.7(C)   All years lived with disability in males per 100,000 lung cancer cases.

out of every 100,000 instances. It is noteworthy that women in India had the lowest burden of disability-adjusted life years, which may indicate improved health outcomes. Conversely, Pakistan and Australia have modest levels of disability-adjusted life years, highlighting the continued necessity of addressing gender-based healthcare inequities in these regions.

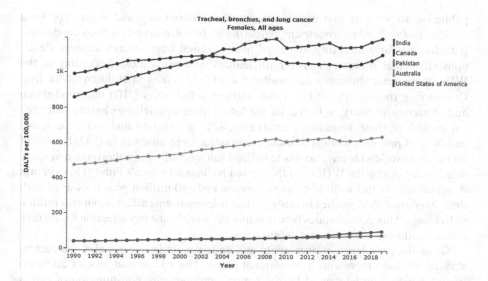

FIGURE 5.7(D) All years lived with disability in females per 100,000 lung cancer cases.

The areas most impacted were the lower reaches of the social and economic spectrum. The three main factors that affected lung cancer mortality and disability-adjusted life years (DALYs), were air pollution, tobacco use, and unhealthful diet. The ubiquitous concentration of fine particulate matter in the atmosphere (PM2.5) contributed weakly to the worsening of lung crevasses, especially in less developed regions [32]. As it is negotiated the terrain of tobacco-related difficulties, a gendered narrative came to light. Up until 2013, women were more vulnerable to costs related to tobacco use, which suggests that quitting smoking was not accomplished as quickly as it may have been for men. This suggests that attempts to restrict tobacco use worldwide have had an impact on women. Nevertheless, concerns about the current route are reflected in the steady decline in the smoker rate. Metabolic risks, as seen by elevated blood sugar levels while fasting, contributed to further complexity [33]. In contrast to the stability of males, the prognosis for women started to improve. The complex relationship between fasting glucose levels and survival chances was shown in a report that emerged from the connections between diabetes and lung cancer mortality. The strength of our investigation resides in providing an extensive assessment of the worldwide burden. However, a warning is there in the data mixture [43].

5.5.5 NUTRITIONAL DEFICIENCIES

The World Health Organization (WHO) defines "micronutrients" as substances that are needed in extremely small amounts, less than 100 mg per day. The creation of hormones, enzymes, and other critical chemicals that regulate growth and development need micronutrients like vitamins and minerals. Many serious issues affect public health, such as iron deficiency, vitamin A deficiency, and goitre. The term "hidden hunger" or "micronutrient malnutrition" refers to several serious conditions, including as early mortality, impaired health, visual impairments, delayed development, mental illnesses, learning difficulties, and exhaustion. According to the WHO, undernourishment is responsible for 45% of deaths in children under five. Considering the significant harm that nutrient deficiencies (ND) due to children and expectant mothers, it is crucial for low-income countries to ensure that they get enough of these nutrients. Remarkably, 42% of children under five years old and 40% of pregnant women in these areas suffer from anaemia [34]. Only 66% of households worldwide have access to iodized salt, according to a thorough investigation carried out by the WHO and the United Nations Children's Fund. This vitamin A deficit affects 19.1 million pregnant women and 190 million preschool-aged children. Moreover, deficiencies in other critical micronutrients affect about two million individuals. This demonstrates how urgently the world must pay attention to and take action to address micronutrient deficiencies.

Given the prevalence of overweight and underweight issues in India, the current state of nutrition represents a transitional phase. The widespread lack of micronutrients, which may be caused by the nation's predominant consumption of cereal products, remains a significant challenge despite these concerns. In 2016 in India, the overall death rate from hunger was 0.5% (95% CI 0.4%, 0.6%), highlighting the severity of the issue. The National Family Health Survey brought to light India's concerning status as the world leader in the burden of anaemia. In 2016, there was a notable increase in the prevalence of anaemia, with 58.6% among children, 53.2% among women who were not pregnant, and 50.4% among older women [35]. It was found that 32% of teenagers and 19% of preschoolers had low zinc levels. Moreover, it was shown that 23% of preschoolers and 37% of teenagers had folate deficits. Significant vitamin deficiencies also occur, with vitamin B12, vitamin A, and vitamin D deficiencies ranging from 14% to 31% in children and adolescents in the preschool age group. Concerning anaemia rates of 67% among preschoolers and 69% among teenagers were found at the same time. In addition, the Goitre prevalence indicated that 3.9% of people had iodine-deficient diseases [36]. The findings of this study emphasize the critical need for an all-encompassing approach to address the various dietary issues facing the Indian population. 5.8 (a) depicts the disease-adapted life for 100,000 cases of malnutrition for all men, illustrating varying pressure levels in different geographic areas. The spectrum of disability in nations like Pakistan and India has decreased noticeably as compared to other locations, showing serious difficulties in controlling sickness and early mortality. Men in the US and Canada also exhibit an unusually high rate of years lived with a handicap per 100,000 instances, however, this disparity is less pronounced than it is for women. India, on the other hand, has the lowest rate of disability-adjusted life per 100,000 cases of women's dietary deficiencies, which may point to improved health outcomes for women. This time, Pakistan and Australia show somewhat adjusted living standards for 100,000 undernourished people, highlighting the ongoing need to tackle gender-based health disparities across all regions.

The World Health Organization and the Ministry of Health and Family Welfare in India conducted a recent study that found variations in the prevalence of anaemia

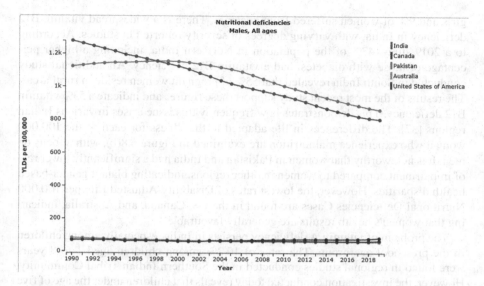

FIGURE 5.8(A) All males; disability-adjusted life per 100,000 nutritional deficiencies cases.

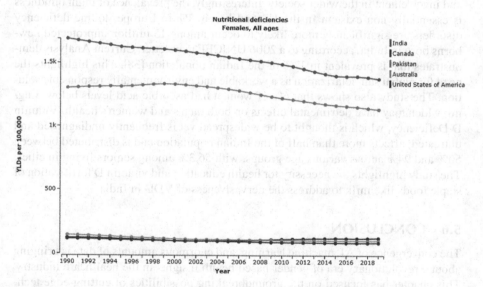

FIGURE 5.8(B) All females; disability-adjusted life per 100,000 nutritional deficiencies cases.

among India's various demographic groups. 58.6%, 53.1%, 50.4%, and 22.7% were reported in the groups of children between 6 and 59 months, women between 15 and 49 years, mothers between 15 and 49 years, and men between 15 and 49 years. A study conducted by the Indian Council of Medical Research revealed that 84% of girls and 9% of women suffered from anaemia. There is a widespread vitamin B12 deficiency in India, with varying degrees of severity reported in studies. According to a 2019 study, 47% of the population in Northern India, including a higher percentage of those with diabetes, had a vitamin B12 deficiency. An additional study conducted in South India revealed that 55% of pregnant women reside in rural areas. The results of the most recent study support these figures and indicate a 53% vitamin B12 deficiency. This demonstrates how frequently this issue arises in various Indian regions [37]. The differences in life adapted to the illness for each of the 100,000 women who experience malnutrition are examined in Figure 5.8(b), with a focus on men. It is noteworthy that women in Pakistan and India had a significantly lower rate of impairment compared to women in other regions, indicating blatant gender-based health disparities. However, the lowest rates of Disability-Adjusted Life per 100,000 Nutritional Deficiencies Cases are found in the US, Canada, and Australia, indicating that women's health results are generally favourable.

The problem of vitamin A deficiency persists in India, especially among children in the preschool age group. The rates of 10.2% among children aged 1 to 8 years were found in regional studies conducted in the Southern Indian Tribal Community. However, the investigation conducted today reveals that children under the age of five have a greater frequency at 19%. This points to potential geographical differences impacted by things like dietary preferences, financial status, mobility for healthcare, and involvement in the wider society. Interestingly, the prevalence of night blindness is essentially non-existent in the US and barely 1% in Europe. Iodine deficiency disorders are significantly more likely to occur among 13 million unprotected newborns born in India, according to a 2006 UNICEF newsletter. Current Analysis demonstrates that is prevalent in 17% of the Indian population [38]. This highlights the need for global salt-solarization as a workable and environmentally responsible solution. The study also shows that 6% of women had ascorbic acid levels below 3 ng/ml, which may have detrimental effects on both men's and women's health. Vitamin D Deficiency, which is thought to be widespread yet is frequently undiagnosed and untreated, affects more than half of the Indian population and is distributed between 50% and 94% across various age groups, with 56.3% among seniors living in cities. The study highlights the necessity for health education and vitamin D fortification of staple foods like milk to address the pervasiveness of VDD in India.

5.6 CONCLUSION

The convergence of AI, machine learning, and enormous amounts of data is bringing about a revolutionary era of gender-based health insights in the healthcare industry. This chapter has focused on the groundbreaking possibilities of cutting-edge technologies and predictive analytics, which provide so far unheard-of opportunities to transform healthcare [41]. Gender-specific health data are becoming a valuable tool that guides the healthcare system toward inclusion and conscious decision-making.

A clarified outline of variations is provided by the investigation of gender-related health issues, such as prostate, ovarian, breast, and lung cancer, and mental health issues [42]. This highlights the urgent need to change paradigms and emphasizes the need to overcome conventional prejudices in research methods.

The accuracy of algorithms plays an important role in aggressively overcoming biases in predictive analytics. The necessity for further improvement is highlighted by the revolutionary potential of AI in observational studies. ML algorithms have a vital role in the fields of predictive medicine and cancer diagnosis. They facilitate the identification and optimization of assisted technologies in high-risk patients. This chapter highlights the use of AI, ML, and predictive analytics. A future where new advancements in technology lead to comprehensive and equitable health outcomes toward gender-inclusive medical care.

REFERENCES

[1] Lounsbury, Olivia, Lily Roberts, Jonathan R. Goodman, Philippa Batey, Lenny Naar, Kelsey M. Flott, Anna Lawrence-Jones, Saira Ghafur, Ara Darzi, and Ana Luisa Neves. "Opening a 'Can of Worms' to explore the public's hopes and fears about health care data sharing: Qualitative study." Journal of Medical Internet Research 23, no. 2 (2021): e22744.

[2] Smith, Benjamin, Anahita Khojandi, and Rama Vasudevan. "Bias in reinforcement learning: A review in healthcare applications." ACM Computing Surveys 56, no. 2 (2023): 1–17.

[3] Nielsen, Mathias W., Marcia L. Stefanick, Diana Peragine, Torsten B. Neilands, John P. A. Ioannidis, Louise Pilote, Judith J. Prochaska et al. "Gender-related variables for health research." Biology of Sex Differences 12 (2021): 1–16.

[4] Kuruvila, Mahima, Eden Estevez, Aruna Anantharaj, and Anjali Mediboina. "Gender equality in antiphospholipid syndrome publications: A comprehensive analysis of first author trends." Cureus 15, no. 12 (2023).

[5] Goldenberg, S. Larry, Guy Nir, and Septimiu E. Salcudean. "A new era: Artificial intelligence and machine learning in prostate cancer." Nature Reviews Urology 16, no. 7 (2019): 391–403.

[6] Jennifer S Temel, Joseph A Greer, Alona Muzikansky, Emily R Gallagher, Sonal Admane, Vicki A Jackson, Constance M Dahlin, Craig D Blinderman, Juliet Jacobsen, William F Pirl, J Andrew Billings, Thomas J Lynch (2010). Early palliative care for patients with metastatic non–small-cell lung cancer. The New England Journal of Medicine, 363(8), 733–742. doi:10.1056/nejmoa1000678

[7] Marie A. Bakitas, Tor D. Tosteson, Zhigang Li, Kathleen D. Lyons, Jay G. Hull, Zhongze Li, J. Nicholas Dionne-Odom, Jennifer Frost, Konstantin H. Dragnev, Mark T. Hegel, Andres Azuero, and Tim A. Ahles. Early versus delayed initiation of concurrent palliative oncology care: Patient outcomes in the ENABLE III randomized controlled trial. Journal of Clinical Oncology 33, no. 13 (2015 May 1): 1438–1445.

[8] Kumar, Vinod and Brijesh Bakariya. "An empirical identification of pulmonary nodules using deep learning." Design Engineering (2021): 13468–13486, https://thedesignengineering.com/index.php/DE/article/view/4610.

[9] Mottet, Nicolas, Joaquim Bellmunt, Michel Bolla, Erik Briers, Marcus G. Cumberbatch, Maria De Santis, Nicola Fossati et al. "EAU-ESTRO-SIOG guidelines on prostate cancer. Part 1: Screening, diagnosis, and local treatment with curative intent." European Urology 71, no. 4 (2017): 618–629.

[10] Kanagasingam, Yogesan, Di Xiao, Janardhan Vignarajan, Amita Preetham, Mei-Ling Tay-Kearney, and Ateev Mehrotra. "Evaluation of artificial intelligence—based grading of diabetic retinopathy in primary care." JAMA Network Open 1, no. 5 (2018): e182665–e182665.

[11] de la Calle, Claire Marie, Hao Gia Nguyen, Ehsan Hosseini-Asl, Clarence So, Richard Socher, Caiming Xiong, Lingru Xue, Peter Carroll, and Matthew R. Cooperberg.

"Artificial intelligence for streamlined immunofluorescence-based biomarker discovery in prostate cancer." Journal of Clinical Oncology 38, no. 6 (suppl) (2020): 279, https://doi.org/10.1200/JCO.2020.38.6_suppl.279.

[12] To, Teresa, Sanja Stanojevic, Ginette Moores, Andrea S. Gershon, Eric D. Bateman, Alvaro A. Cruz, and Louis-Philippe Boulet. "Global asthma prevalence in adults: Findings from the cross-sectional world health survey." BMC Public Health 12, no. 1 (2012): 1–8.

[13] Kaggle Healthcare Dataset. www.kaggle.com/datasets/tomaslui/healthcare-dataset/data.

[14] Fuhlbrigge, Anne L., Benita Jackson, and Rosalind J. Wright. "Gender and asthma." Immunology and Allergy Clinics 22, no. 4 (2002): 753–789.

[15] Eder, Waltraud, Markus J. Ege, and Erika von Mutius. "The asthma epidemic." New England Journal of Medicine 355, no. 21 (2006): 2226–2235.

[16] Gupta, P. D. "Specific needs for healthcare on gender basis in infectious diseases and cancers." International Journal of Clinical Case Reports 2, no. 2 (2023).

[17] Kavousi, Shahin, Najmeh Maharlouei, Alireza Rezvani, Hossein Akbari Aliabad, and Hossein Molavi Vardanjani. "Worldwide association of the gender inequality with the incidence and mortality of cervical, ovarian, endometrial, and breast cancers." SSM-Population Health (2024): 101613.

[18] Sledge, Piper. "From decision to incision: Ideologies of gender in surgical cancer care." Social Science & Medicine 239 (2019): 112550.

[19] Rana, Rezwanul Hasan, Fariha Alam, Khorshed Alam, and Jeff Gow. "Gender-specific differences in care-seeking behaviour among lung cancer patients: A systematic review." Journal of Cancer Research and Clinical Oncology 146 (2020): 1169–1196.

[20] Jabbour, Jana, Merette Khalil, Anna Rita Ronzoni, Ruth Mabry, Ayoub Al-Jawaldeh, Maha El-Adawy, and Hala Sakr. "Malnutrition and gender disparities in the Eastern Mediterranean Region: The need for action." Frontiers in Nutrition 10 (2023): 1113662.

[21] Cirillo, Davide, Silvina Catuara-Solarz, Czuee Morey, Emre Guney, Laia Subirats, Simona Mellino, Annalisa Gigante et al. "Sex and gender differences and biases in artificial intelligence for biomedicine and healthcare." NPJ Digital Medicine 3, no. 1 (2020): 81.

[22] Pavord, Ian D., Richard Beasley, Alvar Agusti, Gary P. Anderson, Elisabeth Bel, Guy Brusselle, Paul Cullinan et al. "After asthma: Redefining airways diseases." The Lancet 391, no. 10118 (2018): 350–400.

[23] Lopes, Sílvia Oliveira, Lívia Carvalho Sette Abrantes, Francilene Maria Azevedo, Núbia de Souza de Morais, Dayane de Castro Morais, Vivian Siqueira Santos Gonçalves, Edimar Aparecida Filomeno Fontes, Sylvia do Carmo Castro Franceschini, and Silvia Eloiza Priore. "Food insecurity and micronutrient deficiency in adults: A systematic review and meta-analysis." Nutrients 15, no. 5 (2023): 1074, https://doi.org/10.3390/nu15051074.

[24] Kulik-Rechberger, Beata, and Magdalena Dubel. "Iron deficiency, iron deficiency anaemia and anaemia of inflammation—an overview." Annals of Agricultural and Environmental Medicine 31, no. 1 (2024): 151–157.

[25] Dicker, Daniel, Grant Nguyen, Degu Abate, Kalkidan Hassen Abate, Solomon M. Abay, Cristiana Abbafati, Nooshin Abbasi et al. "Global, regional, and national age-sex-specific mortality and life expectancy, 1950–2017: A systematic analysis for the Global Burden of Disease Study 2017." The Lancet 392, no. 10159 (2018): 1684–1735.

[26] Stanaway, Jeffrey D., Ashkan Afshin, Emmanuela Gakidou, Stephen S. Lim, Degu Abate, Kalkidan Hassen Abate, Cristiana Abbafati et al. "Global, regional, and national comparative risk assessment of 84 behavioural, environmental and occupational, and metabolic risks or clusters of risks for 195 countries and territories, 1990–2017: A

systematic analysis for the Global Burden of Disease Study 2017." The Lancet 392, no. 10159 (2018): 1923–1994.

[27] Rakhi Dandona, G Anil Kumar, R S Dhaliwal, Mohsen Naghavi, Theo Vos, D K Shukla, Lakshmi Vijayakumar, G Gururaj, J S Thakur, Atul Ambekar, Rajesh Sagar, Megha Arora, Deeksha Bhardwaj, Joy K Chakma, Eliza Dutta, Melissa Furtado, Scott Glenn, Caitlin Hawley, Sarah C Johnson, Tripti Khanna, Michael Kutz, W Cliff Mountjoy-Venning, Pallavi Muraleedharan, Thara Rangaswamy, Chris M Varghese, Mathew Varghese, K Srinath Reddy, Christopher J L Murray, Soumya Swaminathan, and Lalit Dandona. "Gender differentials and state variations in suicide deaths in India: The Global Burden of Disease Study 1990–2016." The Lancet Public Health 3, no. 10 (2018): e478–e489, https://doi.org/10.1016/s2468-2667(18)30138-5.

[28] Jadhav, Apoorva, Abigail Weitzman, and Emily Smith-Greenaway. "Household sanitation facilities and women's risk of non-partner sexual violence in India." BMC Public Health 16 (2016): 1–10.

[29] Sanz-Barbero, Belén, Natalia Barón, and Carmen Vives-Cases. "Prevalence, associated factors and health impact of intimate partner violence against women in different life stages." PLoS One 14, no. 10 (2019): e0221049.

[30] Moitra, Modhurima, Damian Santomauro, Pamela Y. Collins, Theo Vos, Harvey Whiteford, Shekhar Saxena, and Alize J. Ferrari. "The global gap in treatment coverage for major depressive disorder in 84 countries from 2000–2019: A systematic review and Bayesian meta-regression analysis." PLoS Medicine 19, no. 2 (2022): e1003901.

[31] Williamson, Peace Ossom, and Christian I. J. Minter. "Exploring PubMed as a reliable resource for scholarly communications services." Journal of the Medical Library Association: JMLA 107, no. 1 (2019): 16.

[32] Singla, Rajiv, Arpan Garg, Vineet Surana, Sameer Aggarwal, Geetu Gupta, and Sweta Singla. "Vitamin B12 deficiency is endemic in Indian population: A perspective from North India." Indian Journal of Endocrinology and Metabolism 23, no. 2 (2019): 211.

[33] Kumar, Vinod and Brijesh Bakariya. "Classification of malignant lung cancer using deep learning." Journal of Medical Engineering & Technology 45, no. 2 (2021): 85–93, https://doi.org/10.1080/03091902.2020.1853837.

[34] Awotunde, Joseph Bamidele, Abidemi Emmanuel Adeniyi, Sunday Adeola Ajagbe, and Alfonso González-Briones. "Natural computing and unsupervised learning methods in smart healthcare data-centric operations," in Cognitive and Soft Computing Techniques for the Analysis of Healthcare Data, Elsevier, 2022, pp. 165–190.

[35] Braveman, Paula. "Defining health equity." Journal of the National Medical Association 114, no. 6 (2022): 593–600.

[36] Lund, Emily M., and Claire M. Burgess. "Sexual and gender minority health care disparities: Barriers to care and strategies to bridge the gap." Primary Care: Clinics in Office Practice 48, no. 2 (2021): 179–189.

[37] Rajasegaran, Jathushan, Vinoj Jayasundara, Sandaru Jayasekara, Hirunima Jayasekara, Suranga Seneviratne, and Ranga Rodrigo. "Deepcaps: Going deeper with capsule networks," in Proceedings of the IEEE/CVF Conference on Computer Vision and Pattern Recognition, pp. 10725–10733. 2019.

[38] Raftery, Philomena, Natasha Howard, Jennifer Palmer, and Mazeda Hossain. "Gender-Based Violence (GBV) coordination in humanitarian and public health emergencies: A scoping review." Conflict and Health 16, no. 1 (2022): 37.

[39] Prabha, Chander; Singh, Jaspreet; Agarwal, Shweta; Verma, Amit; Sharma, Neha, "Introduction to computational intelligence in healthcare," in Computational Intelligence in Healthcare, Boca Raton: CRC Press, 2022, pp. 1–15.

[40] Verma, Kanupriya et al. Latest tools for data mining and machine learning (Article) (Open Access).

[41] Ramesh TR, Umesh Kumar Lilhore, Poongodi M, Sarita Simaiya, Amandeep Kaur, Mounir Hamdi, "Predictive analysis of heart diseases with Machine Learning approaches," Malaysian Journal Computer Science (2022): 132–148.

[42] Sharma, Geetika, and Chander Prabha, "A systematic review for detecting cancer using machine learning techniques," in Innovations in Computational and Computer Techniques: ICACCT-2021, 2022.

[43] Agarwal, Shweta, and Chander Prabha, "Analysis of lung cancer prediction at an early stage: A systematic review," in Lecture Notes on Data Engineering and Communications Technologies, Singapore: Springer Nature Singapore, 2022, pp. 701–711.

6 Machine Learning's Precision in Tailoring Healthcare Solutions

Shubham Gupta and Harashleen Kour

6.1 INTRODUCTION

Integration of suitable machine learning has a transformative force that has emerged in the realm of healthcare, revolutionising the approach to the deliverance of medical services as well as the treatment of patients. the inclusion of the unparalleled ability to assimilate and evaluate vast portions of data through the identification of patterns, as well as the extraction of valuable observations. The integration of machine learning has allowed a new era of precision medicine. This technology enables healthcare providers to tailor treatment plans and interventions to the specific needs of individual patients, thereby optimising outcomes and improving overall quality of care. The usability of the efficacy of machine learning in the healthcare sector lies in its capacity to analyse diverse data sources through the inclusion of electronic health records, medical imaging scans, genetic information, and wearable devices. The social determinants of health. Through the engagement of synthesising these disparate data streams, machine learning algorithms can assimilate correlations and predict patient outcomes with remarkable accuracy [5]. This empowerment of the predictive power of healthcare professionals to intervene proactively along with the prevention of diseases before they manifest or mitigate their progression more effectively is the prime concern of the report. Association of different elements that can be evaluated according to the norm of the formation of the market allocation of technologies, machine learning facilitates the development of personalised treatment strategies by considering a patient's unique genetic makeup, medical history, lifestyle factors, and environmental influences. The report will engage in the evaluation of all these factors in a systematic manner.

6.2 LITERATURE SURVEY

F. Ali et al. (2020) intended a SHS to predict cardiac disorder depending upon EDP and FF methods [28]. Firstly, the latter method emphasized on integrating the extracted attributes from sensor data and electronic medical records (EMRs) for deriving the valuable healthcare data. Secondly, the information gain (IG) method was employed for removing unrelated and dismissed features, and selecting the significant ones to alleviate the computing burden and enhancing the system efficacy. Lastly, the initial method was employed to train the intended system. This system

DOI: 10.1201/9781003473435-6

predicted the HD at 98.5% accuracy. The experimental outcomes depicted that the intended system was robust to predict heart disease.

Bharadwaj et al. (2021), "A review on the role of machine learning in enabling IoT based healthcare applications" [3], presents a comprehensive review of machine learning applications in IoT-based healthcare, covering domains such as diagnosis, prognosis, monitoring, and logistics.

In the third source, Jayatilake et al. "Involvement of machine learning tools in healthcare decision making" [9] highlighted on role of machine learning in healthcare decision, it plays vital analysis of complex medical data, medical reports, medical images to obtain signals and pass to the algorithm to detect disease at its early stage. Kasula, et al., "Optimization of machine learning algorithms for early detection and risk ranking of prostate cancer" [11]. The fourth research target machine application to detect and make intervention at its earlier stage for chronic disease, through pooled and diverse datasets to predict disease risk and come up with personalised intervention solutions. A. Pawar et al., "Incorporating Explainable Artificial Intelligence XAI to aid the Understanding of Machine Learning in the Healthcare Domain." [17]. The model solves the gap between ML engineers and medical practitioners by making explainability in AI-based systems possible. This system introduces XAI techniques to better understand ML models in the healthcare domain.

Applied includes an algorithm or program that has been made available to healthcare professionals on a public or private platform. It also describes the advanced medical applications used in clinics, hospitals, and other facilities. An experiment consists of an algorithm or application campaigning in a research investigation.

TABLE 6.1
Applications of AI and ML in Healthcare, Both in Present and Future

Healthcare Area	Type of Machine Learning Model	Description	Applied or Experiment	References
EHRs	SVM, DT	Prediction of diagnoses using Electronic Health Records	Applied	Liang et al. 2014 [29]
-	LSTM, CNN	Chronic Disease Prediction from Medical Notes	Experiment	Liu, Zhang & Razavian 2018 [31]
-	RNN	Prediction of post-stroke pneumonia using deep neural networks	Experiment	Ge et al., 2019 [30]
Medical Imaging	CNN	Dermatologist-level classification of skin cancer	Experiment	Esteva et al. 2017 [32]
-	Decision tree	Integration of clinical and imaging features for depression classification	Experiment	Patel et al., 2015 [33]
-	RNNs	Optimized CRISPR guide RNA design for high-fidelity Cas9 variants	Applied	Wang et al., 2019 [34]
-	Random Forest	Identification of high-efficiency target sites for HDR-mediated nucleotide editing	Applied	O'Brien et al., 2019 [35]

6.3   ML INTEGRATION IN THE HEALTHCARE PROVISION AND SERVICES UPLIFTMENT

Machine learning incorporation involves complex initiatives that utilize a variety of techniques in the healthcare field and service elevation. The ML semi-supervised learning model uses labeled and unlabeled information, preeminent to enhancing predictive relevance and manageability of healthcare services [7]. The other element of ML is supervised learning, which focuses on foreboding approaches to patient diagnosis and prediction. This element has been instrumental in aiming to use labeled information to teach algorithms to identify various patterns seen in a patient's medical recording as well as imaging scans. The machine learning unsupervised learning model helps gather patient data for health management and identify trends and inconsistencies without the availability of labeled data in the provision of healthcare.

The last element of ML is the reinforcement learning model, which inputs maximizing treatment plans and resource allocation, through a continuous feedback process that inputs to learn, adopt, and grow healthcare delivery [8]. With integrating the elements of ML, healthcare services are enhanced, and reforms are applied by modernizing processes with statistical precision, customized patient care, and diminishing of errors and mistakes. Healthcare providers can anticipate patients' requirements, interventions input and respond to it adequately, and resources allocations effectively. Additionally, machine learning puts focusing more on continuous learning while disrupting or reforming product or service delivery to cope with solving the complexity of healthcare and inputting efficiency and innovation to the process of its delivery. The integration involves initial intervention, proactive evaluation of healthcare management, and improves stage of repeat patient satisfaction.

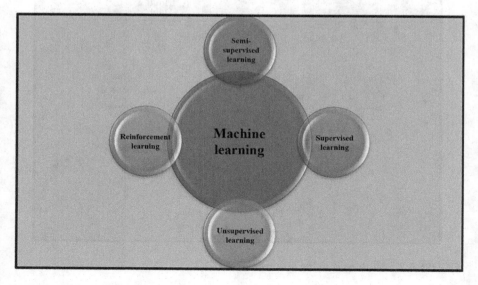

FIGURE 6.1   Machine learning types.

6.4 THE FEATURES OF MACHINE LEARNING FOR HEALTHCARE SERVICES

ML uses several features designed for the sole purpose of meeting different requirements of the healthcare sector that does not fizzle out as services are delivered. ML easily incorporates the EMRS (electronic medical record systems) feature incorporating predictive analytics to enable early disease identification, tailored treatment, and foreseeable patient results from comprehensive patient data. ML models include the features of AI tools that examine a large volume of healthcare information to offer real-time data, enabling clinicians or physicians to make informed decisions, risk evaluation, and customized treatment recommendations [9]. The other integrated aspect of ML is the cloud-data system which facilitates rapid access, secure preservation, and transparency of healthcare information for research, examination, and collaboration among different types of healthcare professionals and providers.

In addition, machine learning incorporated smart unit initiatives by maximizing resource allocation, evaluation of patient flow management, and operational effectiveness inside healthcare services, guaranteeing quality and timely delivery. ML incorporates the features of digital discharge notes which analyze patient data, and treatment plans, as well as provide recommendations for developing communication among healthcare providers. Machine learning-generated smart documents feature which works on efficient documentation procedures, decreasing administrative burden, and compliance activities on healthcare professionals and intensifying

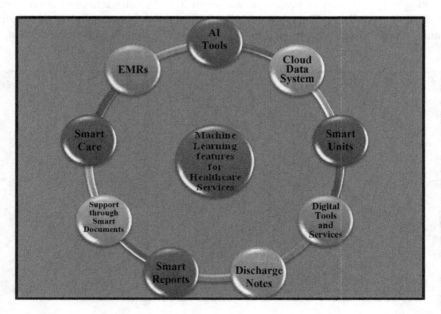

FIGURE 6.2 Machine learning features.

documentation relevancy. Moreover, machine learning incorporates smart report initiatives by examining patient information, outlining key trends, and anticipating analytics to support the healthcare decision-making process, strategies of quality development, and research aspiration [13]. ML intensifies smart care strategies by utilizing anticipating analytics to recognize high-risk associated patients, customize interventions, and observe potential in real-time, encouraging personalized healthcare facilities over time.

6.5 APPLICATION OF ML IN HEALTHCARE

In healthcare services, the application of machine learning transforms different types of features in the sector, starting from predictive analysis to epidemic management, emphasizing more efficient healthcare facilities. ML model examines medical images including MRIs, X-rays, and CT scans to aid in the initial recognition and diagnosis of diseases in the context of fractures, cancer, and abnormalities, intensifying the speed of relevancy and accuracy in the interpretation of medical imaging. ML-associated smart health observation processes regularly investigate patient information from wearable sensors and devices to track crucial signs, identify inconsistencies, and offer customized health information. It influences individuals to effectively maintain their wellness.

Additionally, machine learning recognizes various patterns and biomarkers in patient information to help in disease anticipation, prediction, and risk assessment. It emphasizes initial intervention and obstructive practices to reduce disease development. ML expedites the components of medical research by investigating a large number of datasets to discover new valuable

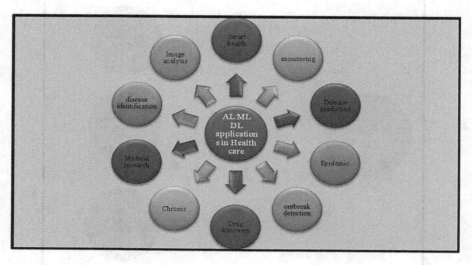

FIGURE 6.3 Application of ML.

interconnections, information, and possible treatments for diseases. It emphasizes customized medicine, drug discovery, and the accuracy of healthcare interventions. Moreover, machine learning incorporated the application of drug discovery platforms which help in the drug improvement procedures by analyzing the interactions of drug targets. It contributes to cost-effective and faster drug discovery methods [3]. ML includes epidemiological data to identify and observe chronic disease outcomes including cardiovascular diseases, diabetes, and infectious diseases, emphasizing initiatives of timely intervention and public health interventions.

6.6   THE IMPACT OF AI AND MACHINE LEARNING IN HEALTHCARE

ML and AI's effect on healthcare is significant, particularly in customized medication, where calculations dissect huge datasets to fit treatments to individual patients, further developing results and lessening negative impacts. NLP (natural language processing) upgrades medical services correspondence by extricating experiences from clinical notes, empowering better navigation and patient consideration coordination. In clinical imaging and diagnostics, AI and ML improve precision and speed in identifying sicknesses from scans, supporting early analysis and therapy arranging [1]. Artificial intelligence and ML alter medical services by upgrading treatments, smoothing out work processes, and saving lives through more exact and effective consideration conveyance.

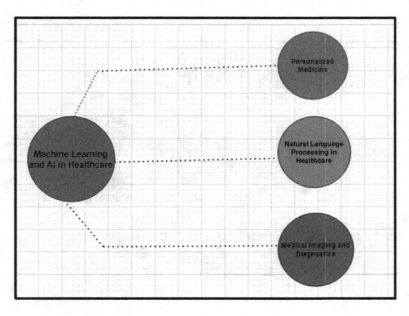

FIGURE 6.4   Personalisation of ML.

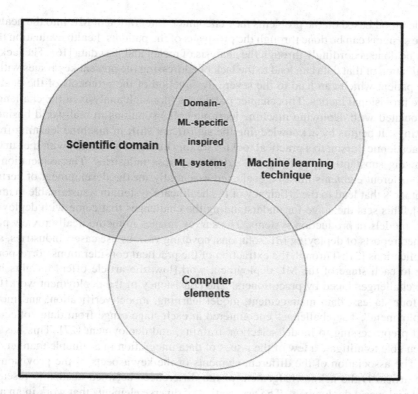

FIGURE 6.5 Elements of ML.

6.7 ML'S SPECIFIC SYSTEMS

In the ML-explicit area, progressions center around optimizing ML frameworks to address explicit difficulties inside scientific spaces. ML strategies are propelled by the complexities of different logical disciplines, directing the advancement of specific calculations and models. These ML frameworks coordinate space explicit information and information to accomplish precise expectations and bits of knowledge. Computer components like processing power, memory, and design assume vital parts in enhancing ML calculations for scientific applications, guaranteeing proficient calculation and examination of complex datasets. The cooperative energy between domain expertise and ML abilities drives advancement and progress across logical spaces, filling disclosures and forward leaps.

6.8 CASE STUDY ANALYSIS

6.8.1 CHALLENGES IN DEPLOYING MACHINE LEARNING: A SURVEY OF CASE STUDIES

The articulation of "Challenges in deploying machine learning: a survey of case studies" highlights the various challenges that healthcare facilities face, and these can be understood based on the problems they encounter. Integration of ML into the health-care systems can be done through the prospects of the patients' health evaluation that can be done accordingly through the analysis of recent and past data [16]. The lack of evaluation of that data can lead to the lack of addressing the present key issues within the patient which can lead to the severe dysfunction of the protocols of the health-care provisional tactics. This chapter provides a thorough analysis of the challenges associated with deploying machine learning (ML) solutions in real-world business settings. It begins by acknowledging the significant shift in machine learning from an academic pursuit to a practical tool for solving business problems, with the survey showing substantial growth in ML adoption across industries. The association of the different elements must be evaluated essentially for the development of certain protocols that lead to the efficiency of the healthcare system in a sustainable manner [14]. This sets the stage for understanding the challenges that come with deploying ML models in production systems. The survey images in methodically review pub-lished reports of deploying ML solutions, covering various use cases, industries, and applications [21]. Through the extraction of the practical considerations correspond-ing to each stage of the ML deployment workflow, the article offers insights into the challenges faced by practitioners. The consistency of the deployment workflow of four stages: "data management, model learning, model verification, and model deployment". The challenges encountered at each stage range from data collection and preprocessing to model selection, training, and deployment [27]. This has not been able to mitigate a few of the issues of data integration in a suitable manner.

The association of the different elements of the key aspects of the provision by the article is the acknowledgement of the differences between academic research and real-world deployment. The association of diverse elements that work in an aca-demic setting may not necessarily translation without any hindrance into practical applications [21]. the formation of the different elements of the Healthcare issues and implementation of ML solutions become more prevalent in industries beyond "Big Tech," the essentiality of understanding the unique challenges along with the complexities for the deployment of these solutions in diverse business environments has been discussed effectively. This can lead to the holistic growth of the study in a systematic manner [2]. The association of different elements of the article also images in the discussion of diverse cross-cutting aspects that impact every stage of the deployment workflow along with the inclusion of ethical considerations, legal compliance, end-user trust, as well as security. These aspects affect the layers of complexity in the deployment process, requiring practitioners to navigate regulatory frameworks, ensure data privacy, and build trust among end-users. Furthermore, the article focuses on the importance of incorporation for addressing challenges at each stage of the deployment process. The recognition that while model learning often acquires the most attention in academic circles, problems such as data management, as well as model verification, are equally crucial for successful deployment. The provision of a comprehensive overview of challenges across the entire deployment workflow, the article lays the groundwork for future research to explore solutions to these issues [6]. The formation of the different analyses of the suitable data accumu-lation as per the requirement of the healthcare sector needs to be The incorporation of the different analyses of case studies as well as the practical experiences adds depth to the survey through the illustration of the real-world examples of challenges encountered by practitioners have been taken into consideration. For example, issues in the collection of data and preprocessing are associated with cases where disparate data sources and data dispersion pose significant hurdles for ML study. Similarly, challenges in model learning, such as model selection and hyper-parameter tuning, are discussed in the context of practical considerations faced by practitioners. The article contributes to a better understanding of the complexities involved in deploying ML solutions in real-world business settings. The establishment of the challenges mitigation tactics has been lacking successively within the study process [22]. Through the systematic analysis of challenges at each stage of the deployment workflow as well as the considering cross-cutting aspects. The provision of valuable observations for practitioners lays the foundation for future research to address these challenges effectively through the discussion within the paper.

6.8.2 Machine Learning Applications for Early Detection and Intervention in Chronic Diseases

The engagement of the suitable research paper is the formation of effective ML applications for the detection of chronic diseases that can be accentuated through the incorporation of the data systems [11]. The indulgence in the association of machine learning (ML) applications has been aimed at the proposition of early detection as well as intervention strategies for chronic diseases [23]. Through recognition of the substantial impact the diversified conditions have on global health can be achieved through the integration of ML. The establishment of the study process is a critical context assimilating the prevalence as well as the burden of chronic diseases, and the implementation of the urgent need for innovative solutions to tackle them. The association of the different elements of the market formation can be done through the engagement of the ML in the evaluation of the disease and symptoms [15]. Through the engagement of the research objectives and inclusion of a thorough exploration of ML applications along with ethical considerations, challenges, as well as the prospects in chronic disease management, the paper associates a comprehensive framework for its investigation. The facilitation of the different elements of the ML integration has been addressed systematically within the paper.

The research provides an examination in a detailed format, of the current landscape by the surrounding ML's integration into early detection and intervention strategies for chronic diseases. The elucidation of the limitations of traditional diagnostic tools and the identification of the necessity of approaches for more sensitive factorisation [24]. The showcasing of the review of the prowess of ML algorithms, specifically for the integration of the predictions of the analytics, in utilisation diverse datasets to forecast disease onset through the enabling of the timely as well as the interventions in a targeted manner [26]. Effectively, the potential of ML for the contribution of personalized healthcare solutions through the analysis of individual health data, a theme recurrent throughout the literature has been highlighted. Ethical considerations along with the emergence of patient privacy have been allocated as a critical focal point within the literature review in accordance with the prompting of the scholars to indulge in issues associated with informed consent along with the responsible use of health data, and the necessity for effective ethical frameworks to indicate ML implementations. Moreover, the review has been able to assimilate the diverse proliferation of various challenges and limitations inherent in ML applications for early disease detection [17][24]. The inclusion of the algorithm bias along with the interpretability, as well as the generalizability of models across diverse populations. This critical examination engages in the underscoring of the different paraphernalia of the importance of ensuring equitable as well as effective ML applications in healthcare.

The integration of the reviewing of the works of literature has been able to anticipate suitable future directions as well as the innovations in ML applications for management of the chronic disease in accordance with "Machine Learning Applications for Early Detection and Intervention in Chronic Diseases." The inclusion of the different integrative principles of emerging technologies like wearable devices and continuous monitoring, alongside advancements in ML algorithms has been done for the enhancement of early detection capabilities [10]. The association of the different elements of the information of the suitable market can be done through the formation of the ML systematic in a sustainable manner. Effective engagement of the healthcare system can lead to the development of the technological factorisation in a sustainable manner. This forward-looking perspective highlights the dynamic nature of research in this field and underscores the potential for ongoing advancements to reshape healthcare practices. This can lead to the suitable focalisation of the technological factors within the systems of healthcare. In conclusion, the literature review essentials elaborate principles of the overview of the current state of research on applications of ML in prospects of the detection as well as the intervention for chronic diseases in an early. The emphasises the different potential of the transformative of ML in revolutionizing practices of healthcare while acknowledging the importance of addressing ethical along with privacy, as well as the technical challenges. The engagement of the setting of the stage for the present study along with the contribution of valuable observations to the discourse on the utilisation of ML for proactive chronic disease management has been done effectively. The review serves as a cornerstone for further research and innovation in this vital area of healthcare.

6.9 RESULTS

The formation of different ideas based on the critical evaluation of the healthcare system at takes through ML configuration and indentation has been systematically taken into consideration in accordance with "Challenges in deploying machine learning: a survey of case studies." This provides a suitable evaluation and observation of the outcomes of employing machine learning (ML) in early detection and intervention strategies for chronic diseases, aligning closely with the overarching thematic engagements [25]. The association of different elements of machine learning can lead to the detection of different strategies which can be applicable for the longevity implementation of the patients in a sustainable manner within the healthcare provision systematics. "The global artificial intelligence (AI) in healthcare market has the size which was roughly estimated at USD 15.1 billion in 2022" [19]. Through the analysis of case studies, implementations, and empirical findings, the section sheds light on the efficacy, challenges, and prospects of ML applications in proactive healthcare management. This can lead to the holistic growth of different criteria of management of the patient in terms of condition and provision of different technological facilitation management in a more diversified manner through the prospect of automated functionalities implementation. This can lead to the holistic growth of the healthcare system it is within the competitive market scenario in case of providing suitable services to patients who need automated services.

The association of different effectiveness of the machine learning algorithms within the identification of different diseases has been taken into account in a systematic manner for the provision of suitable medication and medical facilities under the current needs of the patients. One of the primary results highlighted in the section is the effectiveness of ML algorithms in leveraging diverse datasets to identify subtle patterns and risk factors associated with chronic diseases [18]. Through the employment of predictive analytics models along with the credibility of the researchers, practitioners and other healthcare professionals can forecast disease onset with remarkable accuracy, enabling timely interventions and personalized healthcare solutions. Implementation of this can lead to the holistic growth of different ideas based on the ML services which need to be taken into consideration for providing effective help during critical situations. This finding underscores the. transformative potential of ML in revolutionizing the landscape of healthcare by providing data-driven insights into disease progression and risk assessment. The integration of machine learning which has been "48% between 2017 and 2023," can lead to different applicability of practical factorization which is to be taken into consideration for the formation of different systematics of organisational development in accordance with the requirement of the patients to be handled respective of their conditions [4]. Different management of data can be achieved effectively through the integration of this leading to the betterment of different identification norms of diseases and their chronic and acute levels in a sustainable manner. Moreover, the results reveal the practical implications of integrating ML into chronic disease management. Integration of the steno to the holistic growth or different ideas of individual health data management and approaches of environmental sustainability through lack of wastage of any medical hazardous substances from the hospitals due to the tracking process of machine learning which can enable suitability of the organisation within the market. The engagement of the case studies demonstrates effective algorithmics formats of the ML can analyze individual health data, including genetic, lifestyle, and environmental factors, to customise interventions to specific patient profiles [20]. This personalized approach can holistically optimize the effectiveness of early detection as well as intervention initiatives but also enhance patient outcomes and satisfaction. "23% of healthcare executives in the USA believe that AI and Machine Learning are very effective" has been able to harness the power of ML, and healthcare providers can deliver proactive and targeted care, thereby mitigating the impact of chronic diseases on individuals and healthcare systems.

The formation of different criteria is according to the effective engagement of different stars of data integration and formation can be done through the application of automation of medical billing along with clinical decisions that can support and develop medical practice guidelines in a sustainable manner. However, alongside the promising outcomes, the results also shed light on the challenges and limitations associated with ML applications in healthcare. Implementation of different tactics of prediction along with the diagnosis and prognosis can be developed essentialities t through the maintaining of different strengths of machine learning abilities within the organisation and working process. Ref factors such as algorithm bias along with the interpretability issues, as well as the generalizability of models across diverse populations, increase significant concerns that necessitate careful consideration. Additionally, ethical considerations surrounding patient privacy, informed consent, and the responsible use of sensitive health data pose complex challenges that must be addressed to ensure equitable and effective ML implementations.

The provision of the suitable discourse of the presence of challenges, the results point towards effective management of the implication in the future directions along with the innovations in ML applications for chronic disease management can be evaluated. The effective inclusion of emerging technologies such as wearable devices as well as the implementation of continuous monitoring can lead to the development of ideas for enhancing early detection capabilities and improvement of the patient outcomes. Moreover, advancements in ML algorithms along with the upgradation of the methodologies can facilitate new opportunities

TABLE 6.2
Summary of the Healthcare System

Aspect	Details
Introduction	• Evaluation of healthcare system through ML configuration and indentation
	• Alignment with thematic engagements [25]
	• Global AI in healthcare market size estimated at USD 15.1 billion in 2022 [19]
Efficacy of ML in Healthcare	• ML applications in early detection and intervention for chronic diseases
	• Effectiveness in leveraging diverse datasets [18]
	• Integration of predictive analytics for disease onset forecasting
Integration of ML in Chronic Disease Mgmt	• Practical implications of ML integration
	• Analysis of individual health data for personalized interventions [20]
	• Harnessing ML power for proactive and targeted care
Challenges and Limitations	• Algorithm bias, interpretability issues, generalizability of models
	• Ethical considerations: patient privacy, informed consent, responsible use of health data
Future Directions and Innovations	• Emerging technologies (e.g., wearable devices)
	• Continuous monitoring for enhanced early detection capabilities
	• Advancements in ML algorithms and methodologies for overcoming limitations
Contribution to Proactive Healthcare Mgmt	• Advances understanding of proactive healthcare management
	• Addresses efficacy, challenges, and prospects of ML in healthcare
	• Highlights importance of addressing ethical, privacy, and technical challenges for effective implementations

for overcoming existing limitations and furthering the adoption of proactive healthcare strategies. This can lead to the formation of different ideas of market management in a more sustainable manner. The study engages in the provision of valuable insights into the application of ML in early detection and intervention strategies for chronic diseases. The showcasing of the efficacy, challenges, as well as diverse prospects of ML applications in healthcare has been addressed sustainably [12]. The association of this can lead to the different prospects of the market engagement more sustainably. The section contributes to advancing about the understanding of proactive healthcare management. Through empirical findings and case studies, the results underscore the transformative potential of ML in revolutionizing healthcare practices while highlighting the importance of addressing ethical, privacy, and technical challenges for equitable and effective implementations.

6.10   CONCLUSION

In conclusion, the integration of different system matrices of healthcare through ML algorithmic associations can be done for the betterment of the patient healthcare services provision in a diversified manner. The association of different elements of technological systematics can be derived essentially through the integration of different algorithms which can lead to the betterment of the healthcare service provisions in terms of technical management and different database evaluation for longer periods. Global artificial intelligence in the healthcare market has a substantial share within the systematics of technology factors which need to be developed essentially for the growth of different forecasting tactics of chronic and acute diseases. Association of this kindly to the holistic development of different formative assessments which are needed to be taken into consideration for the deriving of different mitigation practices of the challenges within the machine learning provisions within the healthcare sectors and usability of different functionalities of the technology in a most sustainable manner. This can lead to the betterment of different criteria of management protocol integration of different databases along with privacy security management which has been taken into consideration for the factorization of the patient care systematics. This effectively concludes in a more diversified manner for addressing different key areas of development and usability of ML within the healthcare paraphernalia.

REFERENCES

[1] Ahmed, Z., Mohamed, K., Zeeshan, S. and Dong, X., (2020). Artificial intelligence with multi-functional machine learning platform development for better healthcare and precision medicine. *Database*, 2020, p. baaa010.

[2] Antunes, R.S., André da Costa, C., Küderle, A., Yari, I.A. and Eskofier, B., (2022). Federated learning for healthcare: Systematic review and architecture proposal. ACM Transactions on Intelligent Systems and Technology (TIST), 13(4), pp. 1–23.

[3] Bharadwaj, H.K., Agarwal, A., Chamola, V., Lakkaniga, N.R., Hassija, V., Guizani, M. and Sikdar, B., (2021). A review on the role of machine learning in enabling IoT based healthcare applications. *IEEE Access*, 9, pp. 38859–38890.

[4] Caruana, R., Lundberg, S., Ribeiro, M.T., Nori, H. and Jenkins, S., (2020). August. Intelligible and explainable machine learning: Best practices and practical challenges. In Proceedings of the 26th ACM SIGKDD International Conference on Knowledge Discovery & Data Mining (pp. 3511–3512).

[5] Chen, I.Y., Pierson, E., Rose, S., Joshi, S., Ferryman, K. and Ghassemi, M., (2021). Ethical machine learning in healthcare. Annual Review of Biomedical Data Science, 4, pp. 123–144.

[6] Chua, M., Kim, D., Choi, J., Lee, N.G., Deshpande, V., Schwab, J., Lev, M.H., Gonzalez, R.G., Gee, M.S. and Do, S., (2023). Tackling prediction uncertainty in machine learning for healthcare. Nature Biomedical Engineering, 7(6), pp. 711–718.

[7] Kaur, G., Gupta, M. and Kumar, R., (2021). IoT based smart healthcare monitoring system: A systematic review. Annals of the Romanian Society for Cell Biology, pp. 3721–3728.

[8] Gupta, M., Ahmed, S., Kumar, R. and Altrjman, C. (Eds.), (2023). *Computational Intelligence in Healthcare: Applications,Challenges, and Management*. CRC Press.

[9] Juneja, A., Kumar, R. and Gupta, M., (2022, July). Smart healthcare ecosystems backed by IoT and connected biomedical technologies. In 2022 Fifth International Conference on Computational Intelligence and Communication Technologies (CCICT) (pp. 230–235). IEEE.

[10] Jin, D., Sergeeva, E., Weng, W.H., Chauhan, G. and Szolovits, P., (2022). Explainable deep learning in healthcare: A methodological survey from an attribution view. WIREs Mechanisms of Disease, 14(3), p. e1548.

[11] Kasula, B.Y., (2022). Machine learning applications for early detection and intervention in chronic diseases. International Transactions in Artificial Intelligence, 6(6), pp. 1–7.

[12] Khedkar, S., Gandhi, P., Shinde, G. and Subramanian, V., (2020). Deep learning and explainable AI in healthcare using EHR. In Deep Learning Techniques for Biomedical and Health Informatics (pp. 129–148). Cham: Springer.

[13] Li, W., Chai, Y., Khan, F., Jan, S.R.U., Verma, S., Menon, V.G., Kavita, F. and Li, X., (2021). A comprehensive survey on machine learning-based big data analytics for IoT-enabled smart healthcare system. Mobile Networks and Applications, 26, pp. 234–252.

[14] López-Martínez, F., Núñez-Valdez, E.R., García-Díaz, V. and Bursac, Z., (2020). A case study for a big data and machine learning platform to improve medical decision support in population health management. Algorithms, 13(4), p. 102.

[15] Okay, F.Y., Yıldırım, M. and Özdemir, S., (2021). October. Interpretable machine learning: A case study of healthcare. In 2021 International Symposium on Networks, Computers and Communications (ISNCC) (pp. 1–6). IEEE.

[16] Paleyes, A., Urma, R.G. and Lawrence, N.D., (2022). Challenges in deploying machine learning: a survey of case studies. ACM Computing Surveys, 55(6), pp. 1–29.

[17] Pawar, U., O'Shea, D., Rea, S. and O'Reilly, R., (2020). December. Incorporating Explainable Artificial Intelligence (XAI) to aid the Understanding of Machine Learning in the Healthcare Domain. In Aics (pp. 169–180).

[18] Roscher, R., Bohn, B., Duarte, M.F. and Garcke, J., (2020). Explainable machine learning for scientific insights and discoveries. IEEE Access, 8, pp. 42200–42216.

[19] Saraswat, D., Bhattacharya, P., Verma, A., Prasad, V.K., Tanwar, S., Sharma, G., Bokoro, P.N. and Sharma, R., (2022). Explainable AI for healthcare 5.0: Opportunities and Challenges. IEEE Access, 10, pp. 84486–84517.

[20] Seeliger, A., Pfaff, M. and Krcmar, H., (2019). Semantic web technologies for explainable machine learning models: A literature review. PROFILES/SEMEX@ ISWC, 2465, pp. 1–16.

[21] Seneviratne, M.G., Li, R.C., Schreier, M., Lopez-Martinez, D., Patel, B.S., Yakubovich, A., Kemp, J.B., Loreaux, E., Gamble, P., El-Khoury, K. and Vardoulakis, L., (2022). User-centred design for machine learning in health care: a case study from care management. BMJ Health & Care Informatics, 29(1).

[22] Stiglic, G., Kocbek, P., Fijacko, N., Zitnik, M., Verbert, K. and Cilar, L., (2020). Interpretability of machine learning-based prediction models in healthcare. Wiley Interdisciplinary Reviews: Data Mining and Knowledge Discovery, 10(5), p. e1379.

[23] Vellido, A., (2020). The importance of interpretability and visualization in machine learning for applications in medicine and health care. Neural Computing and Applications, 32(24), pp. 18069–18083.

[24] Vishwarupe, V., Joshi, P.M., Mathias, N., Maheshwari, S., Mhaisalkar, S. and Pawar, V., (2022). Explainable AI and interpretable machine learning: A case study in perspective. Procedia Computer Science, 204, pp. 869–876.

[25] Whig, P., Kouser, S., Bhatia, A.B., Nadikattu, R.R. and Sharma, P., (2023). Explainable machine learning in healthcare. In Explainable Machine Learning for Multimedia Based Healthcare Applications (pp. 77–98). Cham: Springer International Publishing.

[26] Yoon, C.H., Torrance, R. and Scheinerman, N., (2022). Machine learning in medicine: should the pursuit of enhanced interpretability be abandoned?. Journal of Medical Ethics, 48(9), pp. 581–585.

[27] Young, Z. and Steele, R., (2022). Empirical evaluation of performance degradation of machine learning-based predictive models—A case study in healthcare information systems. International Journal of Information Management Data Insights, 2(1), p. 100070.

[28] Ali, F., El-Sappagh, S. and Kwak, K.-S., (2020, 26 June). A smart healthcare monitoring system for heart disease prediction based on ensemble deep learning and feature fusion. Information Fusion, 63, pp. 208–222.

[29] Liang, Z., Zhang, G., Huang, J.X. and Hu, Q.V., (2014). Deep learning for healthcare decision making with EMRs. In Proc—2014 IEEE International Conference on Bioinformatics and Biomedicine (BIBM) (pp. 556–559). Belfast, UK, November 2–5.

[30] Ge, Y., Wang, Q., Wang, L., Wu, H., Peng, C., Wang, J., et al. (2019). Predicting post-stroke pneumonia using deep neural network approaches. International Journal of Medical Informatics, 132, p. 103986. doi:10.1016/j.ijmedinf.2019.103986.

[31] Liu, J., Zhang, Z. and Razavian, N. (2018). Deep EHR: Chronic disease prediction using medical notes. arXiv. 2018:arXiv:1808.04928.

[32] Esteva, A., Kuprel, B., Novoa, R.A., Ko, J., Swetter, S.M., Blau, H.M. and Thrun S., (2017). Dermatologist-level classification of skin cancer with deep neural networks. Nature, 542(7639), pp. 115–118. doi:10.1038/nature21056.

[33] Patel, M.J., Andreescu, C., Price, J.C., Edelman, K.L., Reynolds, C.F. III, and Aizenstein H.J., (2015). Machine learning approaches for integrating clinical and imaging features in late-life depression classification and response prediction. International Journal of Geriatric Psychiatry, 30(10), pp. 1056–1067. doi:10.1002/gps.4262.

[34] Wang, D., Zhang, C., Wang, B., Li, B., Wang, Q., Liu, D., Wang, H., Zhou, Y., Shi, L., Lan, F. and Wang, Y., (2019). Optimized CRISPR guide RNA design for two high-fidelity Cas9 variants by deep learning. Nature Communications, 10(1), p. 4284. doi:10.1038/s41467-019-12281-8.

[35] O'Brien, A.R., Wilson, L.O.W., Burgio, G. and Bauer D.C., (2019). Unlocking HDR-mediated nucleotide editing by identifying high-efficiency target sites using machine learning. Scientific Reports, 9(1), p. 2788. doi:10.1038/s41598-019-39142-0.

7 Real-World Success Stories
How Technology Transforms Gender Healthcare

Palvi Sharma, Rakesh Kumar, and Meenu Gupta

7.1 INTRODUCTION

Wearables, information technology, virtual reality, and the Internet of Things (IoT) have made healthcare industry business and operations transformative. These technological innovations have become increasingly important in our everyday lives. This has led to an era of patient culture in healthcare and more thoughtful caregiving [1]. The history of healthcare has been marked by significant challenges, advancements, and triumphs in gender healthcare. The integration of technology and healthcare has marked a turning point in the history of treatment. The research focuses on the evolving field of gender healthcare, exploring how technological advancements enhance patient outcomes, shape the delivery of healthcare services, and promote inclusion and equity in society [2]. Technological developments are having a profound impact on gender healthcare, and it is crucial to analyse their rapid growth. The innovations, which are shown in Figure 7.1, include telemedicine, wearable health monitoring devices, artificial intelligence and ML applications, electronic health information, data analytics, and mobile health applications. The most important aspect of health promotion is gender-equity in healthcare, which involves the provision of medical services tailored to individual interests and needs [3]. Gender treatment has historically been impeded by gender inequality, discretionary practices, and restrictions, underscoring the need for radical change. With this in mind, the technology seems promising as an effective catalyst to promote and treat long-term gender health problems. The most crucial technological advancements in gender health are telemedicine and remote consultations. Telemedicine is a new type of healthcare that caters to individuals in rural or underprivileged regions where services are not accessible due to poor treatment access. Patients can access care more easily by utilizing virtual consultations with highly qualified physicians, as they are no longer required to travel [4]. Additionally, it allows people to receive high-quality, personalized healthcare through telemedicine at home with little effort on their part. In the context of healthcare, wearable technology refers to devices that people attach to their bodies to collect exercise and health data, which they can then transmit to doctors, insurance companies and other relevant parties. Some examples are blood pressure monitors,

DOI: 10.1201/9781003473435-7

fitness monitors and biosensors [5]. Wearable technologies are smart devices that make our lives easier and help us take better care of our health. These devices can be used to monitor important health indicators such as heart rate, steps taken, sleep quality and even stress levels. Ultimately, this leads to better health and quality of life. In addition, AI and ML tools, as well as telemedicine and wearable health monitoring devices, are changing the field of sexual health. These cutting-edge technologies make it possible to analyse vast amounts of health data, identify patterns and trends, and provide insights that can be used to guide clinical decisions and personalized treatment plans [6]. Many healthcare organizations are utilizing AI and ML to streamline various aspects of their operations, including patient management and administrative tasks. Time is a significant burden for healthcare workers due to paper and administrative tasks. Through the use of artificial intelligence and automation, repetitive tasks can be completed more efficiently, allowing for more time for personal activities and improving patient relationships [7]. Further, the management and delivery of health services that are specific to a particular gender is being modified with the advent of electronic health records (EHR) and data analytics. By utilizing EHRs, medical professionals can streamline paperwork and communicate with their patients, while also enabling seamless care coordination across multiple settings. However, data analysis employs large data sets to identify patterns and potential actions in the population [8].

The technology has the potential to transform gender health, but its implementation is still a challenge. In short, many groups see the rapid development of AI in clinical and biological fields as a great strategy that can support healthcare professionals. Although artificial intelligence has become a significant advancement in the fields of medicine and healthcare, its success has created new questions about medical ethics

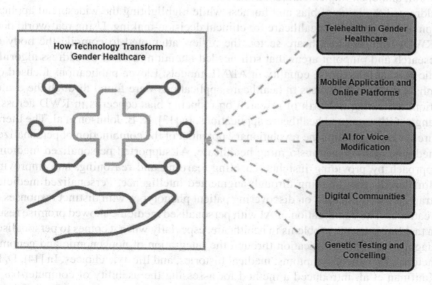

FIGURE 7.1 Modern technology transforms gender healthcare.

[9]. It is therefore important to address these issues and ensure that the technology is used ethically, fairly and in a patient-friendly manner. Finally, future developments in technology and innovation will undoubtedly drive significant progress in the field of gender healthcare. A new era of inclusive, accessible, and high-quality gender-affirming care for all people can be brought in by healthcare providers and policymakers by embracing emerging technologies, encouraging interdisciplinary collaboration, and placing a priority on patient-centric care models. These strategies will help them navigate the challenges presented by the digital age [10].

The chapter is organized as follows: Section 7.2 discusses related work on the transformation of gender healthcare. Section 7.3 introduces the role of technology and success stories in gender healthcare. Section 7.4 highlights the importance of AI and ML in gender healthcare. Challenges and opportunities are discussed in section 7.5. The chapter concludes and proposes future directions in section 7.6.

7.2 BACKGROUND STUDY

This section discusses the research analysis done by different researchers for the transformation of gender healthcare with AI and DL technologies.

In [11], E. Hope Weissler et al. highlighted the ways ML has become more popular as a tool for clinical trials, and the way it can help improve the effectiveness and calibre of biomedical research. Further, the authors also discuss how ML has the potential to increase success rates and lessen human suffering. Cooperation among stakeholders and open communication was essential to properly incorporate ML into clinical research. It was crucial to address issues with data provenance, bias, validation, and regulatory monitoring. For the benefit of all stakeholders, it was imperative that best practices be shared and discussed in order to guarantee the equitable, moral, and open implementation of ML in clinical research. In [12], Y. Huang et al. addressed the issue of bias and fairness while highlighting the widespread application of ML and AI in healthcare for clinical decision-making. Using real-world data (RWD) from the healthcare sector, the review attempted to compile the body of research and pinpoint areas that still needed attention in order to address algorithmic bias and maximise equality in AI/ML models. Eleven publications focused on enhancing model fairness in healthcare applications were found through the evaluation, indicating a dearth of research on reducing bias concerns in RWD across a range of illnesses and healthcare applications. In [13], K. B. Johnson et al. The literature review highlights the revolutionary potential of the combination of personalized medicine and AI in transforming healthcare. AI supported personalized medicine approach by providing insights, enabling learning and reasoning, and improving clinician decision-making through augmented intelligence. Personalized medicine strategies were focused on discovering patient phenotypes with distinct responses to treatment. The combination of AI with personalised medicine showed promise result in tackling intricate problems in healthcare, especially when it comes to personalised diagnosis and prognostication through the integration of nongenomic and genomic factors with patient symptoms, medical histories, and lifestyle choices. In [14], D. R. Kaufman et al. introduced a method for assessing the usability of computer-based healthcare systems that were meant to be used at home by patients. The approach combined novel usability testing techniques with cognitive walkthroughs carried out in patients' homes. The study tried to determine obstacles to the best possible system utilisation as it applied to the IDEATel intervention. The results showed that patient-related issues and interface flaws were impeding performance.

In [15], S. Polevikov et al. highlighted 12 important facets of AI in healthcare, including bias, ethical issues, data protection, and reproducibility. Even though AI has a great deal of potential to improve medical care, there were issues with transparency, bias reduction, and ethical use. To effectively address these difficulties, the research highlights the significance of creating strong best practices in healthcare organisations and promoting multidisciplinary discourse. In [16], T. Chen et al. focused on the recent developments in information technology that have enabled the healthcare industry to gather enormous amounts of big data. This data can be divided into two categories: clinical data, which includes demographics about the patient, medical history, test results, and diagnostic notes; and business-related data, which includes logistical data and operational expenditures. The widespread use of wearable technology make it easier to gather clinical data in real-time, allowing ongoing treatment plan modification and monitoring. In order to maintain consistency, lower errors, improve clinical decision-making, and better comprehend diseases, ML and analytics have become essential tools. These technologies also enabled cost-effective service delivered by revolutionising healthcare management and medical research. In [17], H. Lamba et al. investigated a number of ML approaches in relation to four public policy and social issues. The results showed that these strategies' capacity to increase model fairness is remarkably inconsistent and variable. Post-processing methods, including modifying group-specific score criteria, on the other hand, reliably reduce the differences. These findings highlight the need for more research and the use of post processing techniques to ensure fairness in ML-driven policy applications, which has significant implications for the ML research community as well as practitioners using ML to guide policy decisions. In [18], I. Psychoula et al. proposed a novel LSTM (long short term memory) encoding method that allowed different AAL (ambient assisted living) data views to be created according to the user's access level and the information that was needed. The authors showcased the efficacy and efficiency of the suggested approach with tests conducted on an artificial AAL dataset. Through qualitative demonstration, the authors discussed that how suggested model picks up privacy actions like deletion, disclosure, and generalisation, resulting in nearly flawless data recovery throughout the encoding and decoding phases. In [19], S. Seoni et al. the authors emphasizes the significance of estimating uncertainty in the healthcare industry, especially when it comes to models of AI. This study focused on how DL and ML models in medical can be applied using uncertainty approaches. The review emphasises the necessity of investigating the application of uncertainty approaches to physiological data in future research. Further, including uncertainty methodologies into AI healthcare models can strengthen the dependability of treatment suggestions, increase diagnostic accuracy, and improve decision-making.

In [20], A. D. Amour et al. discussed about the ML systems that when implemented in real-world contexts, they frequently display unexpectedly poor behaviour. The major cause of these failures was the under specification in ML pipelines. The entire process used to train and validate a predictor was called an ML pipeline. When a pipeline can yield numerous unique predictors with similarly high test performance, it was underspecified. By utilising examples from computer vision, medical imaging, natural language processing, clinical risk prediction based on electronic health records, and medical genomics, the authors demonstrated that underspecfication has significant consequences for useful ML pipelines. Further, modelling pipelines for real-world deployment in any sector, under specification must be explicitly taken into consideration. In [21], A. Elhanashi et al. conducted a survey that provides a comprehensive analysis of recent research endeavours, clarifying the designs and uses of several DL models in IoT domains, such as smart cities, healthcare informatics, and surveillance applications. The authors also discussed the future prospects for research, highlighted the necessity for creative methods to get beyond current obstacles and challenges in order to successfully implement DL techniques into IOT frameworks. In [22], S. Hong et al. conducted systematically reviewed the DL techniques used for electrocardiogram (ECG) data. A total of 191 papers from a variety of sources that were published between January 2010 and February 2020 were examined. For applications like disease detection/classification, sleep staging, biometric human identification, annotation/localization, and denoising, a variety of DL architectures has been used. The most successful designs were hybrid ones that combined expert features with convolutional and recurrent neural networks (RNNs). Overall, this systematic analysis highlights obstacles, offers insights into the development of deep learning research employing ECG data, and recommends potential areas of study. In [23], V. S. di Cola et al. investigated the use of knowledge graph (KG) techniques to organise and extract insights from e-health data using health bookings data from Italy's public healthcare system as a case study. This study intended to make it easier to extract medical information and insights by organising the data into a KG and using graph embedding techniques. The authors applied ML approach by using graph embedding to represent entities and characteristics in a single vector space. The findings showed that it was feasible to use KGs in conjunction with supervised and unsupervised ML techniques to examine patients' medical booking patterns. In [24], G. Varoquaux and V. Cheplygina, examined the obstacles restricting the development of computer-aided medical image analysis and makes suggestions for solutions. The study also found systematic problems at every step of technique development and assessment, using evidence from the literature and data challenges. These difficulties include incentives that favour publishing over thorough research, biases in datasets, and constraints in model evaluation. Further, this study highlights continued attempts to tackle these issues in spite of these barriers. It underlines the significance of comprehending the fundamental mechanisms behind these difficulties in order to put certain improvement ideas into practice during the course of the research cycle. In [25], I. Raeesi Vanani and M. Amirhosseini examined how AI and the IoT were integrated in healthcare using disease detection as a focal point. In order to help AI systems detect diseases, the authors describe how IoT devices like wearable monitoring and smart pills collect continuous data. Further, ML techniques were introduced with a focus on the way in which it was used to analyse large amounts of IOT data in order to diagnose problems more quickly.

7.3 ROLE OF TECHNOLOGY IN GENDER HEALTHCARE TRANSFORMATIONS AND SUCCESS STORIES

The use of technology in gender healthcare innovations includes a broad range of applications targeted at enhancing patient outcomes, access, and treatment for people with a variety of gender identities. The following are some significant ways that technology plays a role in this transformation:

7.3.1 TELEHEALTH IN GENDER HEALTHCARE

Telehealth has become a vital tool in the field of gender healthcare, helping to remove obstacles to traditional healthcare services and increase access to care that is gender affirming as shown in Figure 7.2. With this method, gender non-conforming, transgender, and non-binary people can receive gender affirming care without having to physically visit medical institutions. A broad range of services are included in the scope of telehealth in gender healthcare, such as virtual consultations with medical professionals who specialise in transgender healthcare, mental health counselling, hormone therapy management, post-operative care advice, and continuing follow-up appointments [26]–[27]. Some of the advantages to telehealth in gender-affirming care include:

- Enhanced Access: People in underserved or rural areas can communicate remotely with specialised healthcare experts through telehealth, which removes barriers related to geography. Now that gender-affirming treatment is becoming more accessible, transgender and non-binary people can access it regardless of their gender.
- Convenience: The use of telehealth sessions instead of visiting the clinic can lead to time savings and less logistical challenges for patients. The assurance of this comfort guarantees a more precise appointment and better continuity of care.
- Care That Is Culturally Competent: The training provided to telehealth professionals can help them ensure that their care is sensitive to the specific needs and experiences of transgender and non-binary individuals, while also being culturally appropriate. Thus, patients can expect to receive better care with improved and positive outcomes.
- Enhanced Safety and Privacy: Traditional care for transgender people is problematic due to concerns about discrimination and privacy. Patients can receive treatment at home in a more secure and private setting through telehealth.

7.3.2 MOBILE APPLICATIONS FOR HORMONE REGULATIONS

The focus of this section is on how digital health applications and online resources can help meet people's mental health support needs. The following ideas are highlighted by it.

FIGURE 7.2   Telehealth in gender healthcare.

7.3.2.1   Role of Mobile Apps and Online Platforms

The availability of mobile apps and online platforms, which are tailored to their specific needs, is crucial in providing mental health support to transgender and non-binary individuals as shown in Figure 7.3. These digital platforms offer on-demand support anytime, anywhere via computers, cell phones, or tablets, and serve as a life-line for people who might encounter obstacles while attempting to contact traditional mental healthcare services. These platforms' anonymity and privacy are especially beneficial since they let people ask for assistance and talk about their experiences without worrying about prejudice or judgement [28]—[29].

7.3.2.2   Services and Features Offered

The features and services provided by online mental health support platforms and digital health apps include:

- Therapy sessions: Availability of certified counsellors or therapists with expertise in transgender and non-binary mental health.
- Peer Support Network: Online groups where people can interact with others who have gone through similar things and provide support to one another.
- Tools for mental wellness: Apps like mood monitors, breathing techniques, and coping mechanisms designed to meet the particular mental health requirements of people of different gender identities.
- Educational Resources: Resources and information on gender identity, mental health, dealing with prejudice, and getting access to affirming medical treatment.

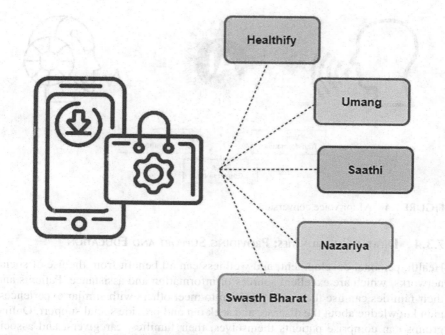

FIGURE 7.3 Role of mobile apps and online platforms.

7.3.3 AI FOR VOICE MODIFICATIONS

For the modification of voices (as shown in Figure 7.4), AI technology plays an important role to change or personalise someone's voice. The AI software can be used for a variety of applications, such as voice synthesis for people with speech impairments, transgender voice training, and voice cloning for voiceovers or digital assistants for entertainment purposes. Here is an outline of how AI is applied to voice modifications:

- Voice Transformation: To accomplish desired results, AI algorithms are able to alter a person's voice's pitch, tone, and other features. AI can help transgender people receiving voice training by either masculinizing or feminising their vocals to better reflect their gender identification.
- Speech Synthesis: The creation of human voices is achieved through speech synthesis technology. The main feature that also led to its creation is the ability to automatically convert typed text into speech.
- Voice Conversion: The process of voice conversion involves conveying the essence or personality of a speaker to another speaker without altering their speech structure.
- Privacy and Security Considerations: By modifying the vulnerability of AI-based health systems during their development, they can be more secure and privacy-protected. Secure voice authentication methods, encryption, and segregated privacy are among the technologies that safeguard user voice data from misuse or illegal use.

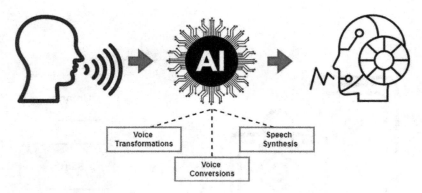

FIGURE 7.4 AI for voice conversion.

7.3.4 DIGITAL COMMUNITIES: PROVIDING SUPPORT AND EDUCATION

Health, personal development, and wellness can all benefit from the use of social networks, which are excellent sources of information and assistance. Patients and their families can use them as a platform to meet others with similar experiences, gain knowledge about the disease, and seek out and provide social support. Online groups can comprise patients themselves, their families, caregivers, and associations of medical professionals with shared interests. As the healthcare system transforms and the internet becomes more widely used, it is being utilized for diseases and health education. Online health communities are everywhere these days, and they have a huge impact on the lives of their members. Despite the lack of consensus on best practices for creating virtual health communities, there is limited research on their effectiveness. The benefits of online health communities are not well understood in current research, but they do appear to be beneficial to both patients and doctors. Webinars, online courses, and workshops are available through digital communities to help people take responsibility for their health and talk about their needs.

7.3.5 GENETIC TESTING AND COUNSELLING

As part of the healthcare team, genetic counsellors provide information and support to families who are at risk for a genetic condition. Their expertise in genetic disorders makes them a crucial resource for patients, the general public, and other medical professionals. To identify families with potential genetic issues, genetic counsellors gather information from family history, inheritance patterns, and the likelihood of recurrence. They offer information on genetic testing and associated procedures. They are trained to provide patients and families with intricate and difficult-to-understand information on genetic risks, testing, and diagnosis. The use of genetic testing and counselling is a vital aspect of human health in medicine, as it provides insight into an individual's genetic profile and potential health issues. Individuals and families can benefit from genetic counselling's personalized guidance, education, and support in comprehending the intricate details revealed by genetic test results. Through the use of genetic testing and counselling services, healthcare providers can empower individuals to make informed choices about their health, improve healthcare delivery, and enhance overall well-being.

7.4 AI AND ML IN GENDER HEALTHCARE

As shown in Figure 7.5, AI and ML technologies are used to solve different kind of issues that occur in gender healthcare. These technologies enhance the care and support that is provided to patients [30]—[31]. In this case, AI and ML are used as follows:

- Customised Treatment Plans: To create individualised treatment plans that are specific to each patient's gender-specific healthcare needs, AI and ML algorithms examine large datasets, including patient demographics, medical histories, genetic data, and therapy reactions. With this method, patients are guaranteed to receive care that is compatible with their particular health needs, preferences, and identities.
- Predictive Modelling: In healthcare, predictive models that are constructed using AI and ML techniques evaluate patient data to estimate treatment responses, detect risk factors, and forecast health outcomes. Healthcare professional's uses past historical data and advanced analytics to suggest better treatment option for patients.
- Diagnostic Support: Different ML algorithms are used to analyse the patient symptoms. These algorithms are trained with the patient's data and after the training it can identify and diagnose gender health problems at an early stage. These AI-based diagnostic tools help the healthcare professionals to make quick decisions and improve the quality of care for patient.
- Research and Innovation: The use of ML and AI to analyze large data sets enables the development of gender-neutral health research and innovation by identifying patterns and providing insights into gender differences in health, treatment outcomes, and epidemiological trends. These developments will reinforce the importance of evidence-based decision-making, advance research excellence, and contribute to the progress of gender-sensitive care.
- Remote Monitoring and Telemedicine: Healthcare can benefit from the use of AI-powered telemedicine systems and remote monitoring tools, which can facilitate virtual consultations, remote assessment, and continuous patient monitoring. With the help of these technologies patients can receive care services from anywhere through the online services.

Gender-nonconforming people's unique needs and preferences are fulfilled through personalized, data driven, and easy-to-use services enabled by AI and ML, ultimately improving quality of life (QOL) and health outcomes.

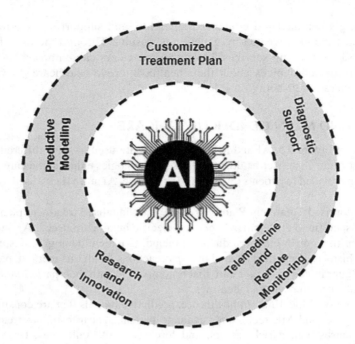

FIGURE 7.5   AI & ML in healthcare.

7.5   CHALLENGES AND OPPORTUNITIES OF USING TECHNOLOGY IN GENDER HEALTHCARE

The role of technology in gender health involves several challenges and opportunities that need to be addressed in order to provide equitable access and high-quality care to patients of all genders. Here we discuss some of the key barriers and opportunities for integrating gender-responsive technologies into healthcare:

7.5.1   CHALLENGES

- The elderly and marginalized groups may face challenges in accessing technology-enhanced healthcare due to inadequate digital health literacy. Education awareness can be helpful for patients.
- To protect sensitive information and maintain patient trust, gender health technology (EHR, telemedicine, apps) must address privacy and security considerations.
- The inclusion of insurance, geographic location, and bias in equitable health technology is necessary to address access disparities among disadvantaged populations.
- To reach underserved populations, gender-equitable health technology must overcome structural barriers such as insurance, geography, and biases.

7.5.2 OPPORTUNITIES

- Through online forums and communities, technology reduces social isolation and helps people navigating gender identity make connections and receive assistance.
- By identifying patterns, discrepancies, and areas that require intervention, big data analytics in gender healthcare help optimise resource allocation and advance health equality initiatives.
- Telemedicine increases the possibilities of receiving healthcare, which strengthens equality, especially in rural areas, through remote consultation and reducing travel barriers.
- The use of technology enhances patient outcomes through the provision of personalized treatments and gender-specific treatments that enhance patient satisfaction.

7.6 CONCLUSION AND FUTURE DIRECTIONS

Applying technology to reduce injustices, improve patient-centered care, and advance health equity is a revolutionary opportunity in gender healthcare. Through the utilization of innovative methods and the promotion of cooperation among interested parties, we may effectively manage obstacles and unleash the complete capacity of technology to enhance the well-being of people with varying gender identities. Further, investigating cutting-edge technologies like wearables, genomics, and AI has the potential to advance gender healthcare and open the door to more specialized and successful therapies. In addition, more investigation into digital health platforms, remote monitoring, and telemedicine is probably going to result in better outcomes and access for a wider range of patient populations.

REFERENCES

[1] S. L. Dworkin, P. J. Fleming, and C. J. Colvin, "The promises and limitations of gender-transformative health programming with men: critical reflections from the field," *Cult. Heal. Sex.,* vol. 17, pp. 128–143, 2015, doi: 10.1080/13691058.2015.1035751.

[2] A. S. George, R. Morgan, E. Larson, and A. Lefevre, "Gender dynamics in digital health: Overcoming blind spots and biases to seize opportunities and responsibilities for transformative health systems," *J. Public Heal. (United Kingdom),* vol. 40, pp. II6–II11, 2018, doi: 10.1093/pubmed/fdy180.

[3] M. R. Kauth, and J. C. Shipherd, "Transforming a system: Improving patient-centered care for sexual and gender minority veterans," *LGBT Health,* vol. 3, no. 3, pp. 177–179, 2016. Mary Ann Liebert, Inc. 140 Huguenot Street, 3rd Floor New Rochelle, NY 10801 USA.

[4] O. Udenigwe, O. Omonaiye, and S. Yaya, "Gender transformative approaches in mHealth for maternal healthcare in sub-Saharan Africa: A systematic review," *Front. Digit. Heal.,* vol. 5, p. 1263488, 2023.

[5] M. J. Neubert, and L. D. Palmer, "Emergence of women in healthcare leadership: Transforming the impact of gender differences," *J. Men's Heal. Gend.,* vol. 1, no. 4, pp. 383–387, 2004.

[6] M. C. Burke, *Transforming gender: Medicine, body politics, and the transgender rights movement*. University of Connecticut, 2010.

[7] W. A. Bapolisi *et al.*, "Impact of a complex gender-transformative intervention on maternal and child health outcomes in the eastern Democratic Republic of Congo: protocol of a longitudinal parallel mixed-methods study," *BMC Public Health*, vol. 20, pp. 1–11, 2020.

[8] K. Doyle, J. Kato-Wallace, S. Kazimbaya, and G. Barker, "Transforming gender roles in domestic and caregiving work: Preliminary findings from engaging fathers in maternal, newborn, and child health in Rwanda," *Gend. & Dev.*, vol. 22, no. 3, pp. 515–531, 2014.

[9] A. Stewart and J. Lander, "Transforming gender relations in an ageing world," Working or Discussion Paper, Faculty of Social Sciences, School of Law, University of Warwick, Coventry, London, pp. 1–18, 2018. Available: https://wrap.warwick.ac.uk/120201/.

[10] N. Jones, R. Holmes, E. Presler-Marshall, and M. Stavropoulou, "Transforming gender constraints in the agricultural sector: The potential of social protection programmes," *Glob. Food Sec.*, vol. 12, pp. 89–95, 2017.

[11] E. H. Weissler *et al.*, "Correction to: The role of machine learning in clinical research: Transforming the future of evidence generation (Trials, (2021), 22, 1, (537), 10.1186/s13063–021–05489-x)," *Trials*, vol. 22, no. 1, pp. 1–15, 2021, doi: 10.1186/s13063-021-05571-4.

[12] Y. Huang *et al.*, "A scoping review of fair machine learning techniques when using real-world data," *J. Biomed. Inform.*, vol. 151, October 2023, p. 104622, 2024, doi: 10.1016/j.jbi.2024.104622.

[13] K. B. Johnson *et al.*, "Precision medicine, AI, and the future of personalized health care," *Clin. Transl. Sci.*, vol. 14, no. 1, pp. 86–93, 2021, doi: 10.1111/cts.12884.

[14] D. R. Kaufman *et al.*, "Usability in the real world: Assessing medical information technologies in patients' homes," *J. Biomed. Inform.*, vol. 36, no. 1–2, pp. 45–60, 2003, doi: 10.1016/S1532–0464(03)00056-X.

[15] S. Polevikov, "Advancing AI in healthcare: A comprehensive review of best practices," *Clin. Chim. Acta*, p. 117519, 2023.

[16] T. Chen, E. Keravnou-Papailiou, and G. Antoniou, "Medical analytics for healthcare intelligence—recent advances and future directions," *Artif. Intell. Med.*, vol. 112, 2021, doi: 10.1016/j.artmed.2021.102009.

[17] H. Lamba, K. T. Rodolfa, and R. Ghani, "An empirical comparison of bias reduction methods on real-world problems in high-stakes policy settings," *ACM SIGKDD Explor. Newsl.*, vol. 23, no. 1, pp. 69–85, 2021.

[18] I. Psychoula *et al.*, "A deep learning approach for privacy preservation in assisted living," in *2018 IEEE international conference on pervasive computing and communications workshops (PerCom Workshops)*, 2018, pp. 710–715.

[19] S. Seoni, V. Jahmunah, M. Salvi, P. D. Barua, F. Molinari, and U. R. Acharya, "Application of uncertainty quantification to artificial intelligence in healthcare: A review of last decade (2013–2023)," *Comput. Biol. Med.*, vol. 165, August, p. 107441, 2023, doi: 10.1016/j.compbiomed.2023.107441.

[20] A. D'Amour *et al.*, "Underspecification presents challenges for credibility in modern machine learning," *J. Mach. Learn. Res.*, vol. 23, no. 226, pp. 1–61, 2022.

[21] A. Elhanashi, P. Dini, S. Saponara, and Q. Zheng, "Integration of deep learning into the IoT: A survey of techniques and challenges for real-world applications," *Electronics*, vol. 12, no. 24, p. 4925, 2023.

[22] S. Hong, Y. Zhou, J. Shang, C. Xiao, and J. Sun, "Opportunities and challenges of deep learning methods for electrocardiogram data: A systematic review," *Comput. Biol. Med.*, vol. 122, p. 103801, 2020.

[23] V. S. di Cola, D. Chiaro, E. Prezioso, S. Izzo, and F. Giampaolo, "Insight extraction from e-Health bookings by means of hypergraph and machine learning," *IEEE J. Biomed. Heal. Informatics,* 2023.

[24] G. Varoquaux, and V. Cheplygina, "Machine learning for medical imaging: Methodological failures and recommendations for the future," *NPJ Digit. Med.,* vol. 5, no. 1, p. 48, 2022.

[25] I. Raeesi Vanani, and M. Amirhosseini, "IoT-based diseases prediction and diagnosis system for healthcare," *Internet Things Healthc. Technol.,* pp. 21–48, 2021.

[26] S. Rahman, S. Amit, and A.-A. Kafy, "Gender disparity in telehealth usage in Bangladesh during COVID-19," *SSM-Mental Heal.,* vol. 2, p. 100054, 2022.

[27] K. M. Goldstein *et al.,* "Telehealth interventions designed for women: An evidence map," *J. Gen. Intern. Med.,* vol. 33, pp. 2191–2200, 2018.

[28] V. Bravou, A. M. S. Driga, and A. Drigas, "Emotion regulation, the function of stress hormones & digital technologies," *Tech. Biochem.,* vol. 3, no. 2, pp. 27–34, 2022.

[29] S. Rahayu, Y. R. Dewahrani, A. Nurkhofiyya, and R. H. Ristanto, "Scaffolding self-regulated learning through Android-based mobile media on hormone system,"*AIP Confe. Proc.,* vol. 2331, no. 1, 2021.

[30] T. G. García-Micó, and M. Laukyte, "Gender, Health, and AI: How Using AI to Empower Women Could Positively Impact the Sustainable Development Goals," in *The Ethics of Artificial Intelligence for the Sustainable Development Goals,* Springer, 2023, pp. 291–304.

[31] N. Buslón, A. Cortés, S. Catuara-Solarz, D. Cirillo, and M. J. Rementeria, "Raising awareness of sex and gender bias in artificial intelligence and health," *Front. Glob. Women's Heal.,* vol. 4, 2023.

8 Patient-Centric Technology Empowerment

Fathimathul Rajeena P.P. and Shakeel Ahmed

8.1 INTRODUCTION

Patient engagement or empowerment has become a key component of patient-centred care. The patients can regulate their health and be more active in their treatments [1–2]. Patient empowerment improves well-being, health status, self-management, and quality of life cost-effectively [3]. Professionals can better treat and educate each patient on an individual basis by identifying and measuring patient empowerment.

Patient empowerment encompasses the following [4]:

- Psychological empowerment
- Community empowerment
- Organisational empowerment

People develop or are provided opportunities to influence their own lives through psychological empowerment, which includes interactional, behavioural, and intrapersonal components [5]. Organisational empowerment consists of both internal activities that provide people with more authority inside the organisation and external mechanisms that allow the organisation to impact choices and policies in the community at large [6]. Community empowerment involves people and groups working together to improve community living and transform the social system. Patients' capacity, power, and knowledge to manage their conditions and lifestyles are integrated into the definitions of patient empowerment. Patients' empowerment must be done by themselves, while healthcare experts can only help [7].

Empowering patients means letting them handle their care. Patients have more power over health decisions and behaviors in a respectful and trusting provider—patient relationship. Internet tools, therapy, and funding information help patients comprehend prognoses, appropriate treatment alternatives, and treatment costs. Empowered patients educate themselves, ask questions, and make decisions about their treatment. Patient empowerment is vital to a high-performing, cost-efficient healthcare system.

8.2 STRATEGIES FOR ATTAINING PATIENT EMPOWERMENT

Strategies for patient empowerment include:

DOI: 10.1201/9781003473435-8

8.2.1 PATIENT-CENTRIC APPROACH

The change from paternalistic to patient-centred treatment is vital to patient involvement. Healthcare practitioners made decisions without considering patient preferences or values in the paternalistic approach. Despite its good intentions, this strategy sometimes results in treatment programs that don't meet patient requirements.

Instead, a patient-centred approach puts the patient in charge of their treatment. Participants collaborate and make decisions together under this paradigm, with clinicians acting as facilitators and guides. A patient-centred approach recognises patients' skills from their life experiences and values. This approach changes people's attitudes towards managing their health. Patients feel more responsibility over their treatment when they are heard, respected, and included in decision-making. Ownership encourages treatment adherence and improves health outcomes.

Healthcare professionals need to be more patient-centred. It requires openness to acknowledge patients' significant thoughts and opinions. It also requires clear and consistent communication that can help patients understand their health issues, treatment options, and risks and benefits.

8.2.2 ELICITING PATIENT PREFERENCES

Finding patient preferences and goals demands attentive listening and open-ended questions. Providers should inspire patients to discuss their health issues and goals. Listening to patients help doctors understand their motives, objectives, and constraints. After that clinicians can collaborate on treatment alternatives. This approach offers clinically sound treatment programs matching patients' expectations and thus, improving adherence and health outcomes.

Increasing the patient's involvement, in issues ranging from treatment options to disease management, is one of the most important challenges. The transition from being mere beneficiaries to fully engaged contributors is critical in cultivating ownership and accountability. For this, patients need information, skills, and support which can be given by their healthcare professionals through training on drug adherence, symptom management, good living habits, and problems.

Beyond the data, professionals, patients, and caregivers having a joint thought about the wisdom of making patients more self-sufficient will be all that is needed. It provides their ability check and also non-manipulative environment, which ensures that patients have an opportunity and are given a listening as they express their cares and are helped. It is the function of the elements formulating the support team for the people who are self-directed learners to ensure that there is a constant communication and cooperation among themselves. This assistance system is required because the process of self-managing its own experiences is quite a long-term activity and, in this case, it should be overseen.

8.2.3 ENCOURAGE PATIENT ENGAGEMENT IN HEALTHCARE SYSTEM TRANSFORMATION

A quick shift is advised in healthcare from the conventional provider-focused models to the patient-driven system which entails inclusion of patients in courses of medical action. It is necessary to promote patients' engagement in the efforts of quality improvement and healthcare system design so that a degree of trust is built to lead to an assurance that patients' views and opinions are acknowledged. It can lead to the positive transformation of patient-precision medication which can be more success-ful and patient-focused.

Patient-centred environment is an important element for health system design to incorporate the input of patients at all stages. Through providing good advice to patients, doctors and caregivers may make patients someone who takes care of himself. This combined and entire approach promotes a sense of ownership and participation, which in turn leads to improved treatment adherence, fair use of health services and well-being.

8.2.4 UTILISE TECHNOLOGY TO IMPROVE PATIENT ENGAGEMENT

Technology has become the major promoter for today's healthcare professionals to provide patients with a broad assortment of tools and assistance to assess their over-all health and to manage it properly. Patient portals, secure messaging systems, and telemedicine platforms facilitate this communication in a more simple, efficient way between providers and patients. This approach can enable a user to pull out their records, arrange for an appointment conveniently, talk to a primary care provider with ease, and so much more, all within the confines of their houses. It becomes much more convenient, especially for the patients with illnesses or limited mobility, to easily reach the doctor and have medical visits. Moreover, since the obstacles to receiving the medical care are removed and the interventions are facilitated, this type of technology becomes very useful.

Technology can also help people get individualised healthcare information and education. Health-related educational videos, quizzes, and mobile applications can be made to fit particular conditions, giving patients the power to learn about their diagnoses, treatment choices, and ways to take care of themselves. This readily available information can aid decision-making and treatment adherence.

8.3 DIGITAL INNOVATIONS FOR PATIENT EMPOWERMENT

Healthcare providers and patients benefit from digital health solutions. The solu-tion or service should put patients—usually end users—at the centre and provide healthcare to everyone, anytime and anywhere, to fully benefit from new technolo-gies. Patient acceptance is essential for digital health products for vulnerable pop-ulations like older adults or mental health support patients [8]. Thus, people may work with doctors and engineers to create digital health solutions that suit them. Patient-centred, interactive qualitative research can help design digital health tech-nologies with users in mind rather than only for them. By providing comprehensive, high-quality treatment, the innovative solutions simplify healthcare in industrialised nations and improve it in developing countries. It replaces doctor-centred care with patient-centred care by electronically gathering and using patient data.

In this chapter, we look into the digital technologies of the fourth Industrial Revolution that enhanced and enabled patient-centric services and treatments in healthcare, aiming for patient empowerment and a few instances from real-world usage.

8.3.1 TELEMEDICINE AND REMOTE HEALTHCARE

Remotely providing medical services, including examinations and consultations, is known as telemedicine, often referred to as telehealth or e-medicine. Telemedicine allows doctors to diagnose and treat patients remotely. A patient visits the nearby local health centre, where a local healthcare professional performs a primary health check-up. The regional healthcare unit collects vital statistics and sends the data to a city hospital. The medical practitioner at that remote hospital then conducts a live interaction with the patient, using audio or video conferencing systems and automated live feeds. The hospitals store the patient data, details, and recorded interactions in the shared database, which can be accessed using mobile apps or a web-based interface. The leading hospitals and the specialist hospitals are also linked for any specialised support needed in case of emergencies. The leading hospitals also have the same teleconferencing units to support remote patients. This system allows for more convenient and efficient healthcare services for patients. Patients can use personal technology or telehealth kiosks to connect with physicians from home, as shown in Figure 8.1. It enables real-time doctor-patient communication [9], [10].

Telemedicine uses telecommunication technology to deliver medical diagnosis, intervention, and therapy [11]. Telehealth also covers remote patient monitoring and medical training services that are non-clinical. Millions of people rely on telemedicine for health promotion and disease prevention. Healthcare professionals use telehealth apps to diagnose and treat patients [12–15].

Telehealth empowers patients by increasing access to care, removing geographical restrictions, enabling remote or limited mobility patients to contact healthcare providers. Telehealth consultations can improve patient satisfaction and reduce the travel burden for numerous diseases. Patients can make appointments, get consultations, and acquire medications from home, increasing convenience. This convenience improves healthcare use and treatment adherence. Telehealth platforms offer

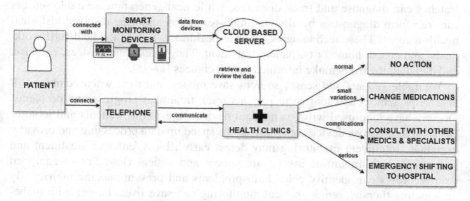

FIGURE 8.1 Digital health monitoring architecture.

educational information and tools to help patients manage chronic diseases between visits, thereby encouraging self-management. Diabetes patients can utilise telehealth platforms to measure their blood sugar, document their food consumption, and receive individualised coaching from doctors [16].

Benefits

- Better access to care for rural and underserved areas
- More consistent care through regular contact and convenient for patients
- Cost savings from less travel
- Flexible appointment times and locations
- Access to specialists not locally available
- Improved communication with providers
- Better coordination of care with access to medical records and reduced waiting times

Challenges

- Limited evidence of effectiveness
- Lack of regulation
- Security and privacy concern
- Technical issues
- Limited access to technology
- Inequitable access based on affordability
- Geographic, time zone, and language barriers
- Social isolation

8.3.2 SENSORS, WEARABLES AND IoTs

Wearable devices use physical, chemical, and biological sensors to collect physiological information in real-time and non-invasively. These devices can be utilised in various ways, such as glasses, jewellery, face masks, wristwatches, fitness bands, tattoo-like devices, bandages, and fabrics (Figure 8.2). Biophysical signals from smartwatches can diagnose and track disorders, while next-generation wearable sensors can transform diagnostics by allowing high-resolution and time-resolved historical health records. These technologies enhance healthcare, reduce medical staff workloads, and enable home or outpatient treatment. They track weight, sleep, exercise, and diet, helping users make informed health choices [17–18].

Wearable gadgets and sensor sources save money and time when combined. It also helps seniors and long-term patients track treatment progress and problems. Doctors have long used wireless medical devices to examine patients and monitor essential data. Cross-device communications speed up data processing and provider exchange. Intelligent monitoring may detect early illness, enhance treatment and prevention, and streamline asset management and patient flow. Sensor-equipped data recorders can identify cold chain problems and prevent vaccine injuries. By anticipating therapy, remote patient monitoring can save lives. People with diabetes may check glucose levels in real-time with continuous glucose monitoring, and

FIGURE 8.2 Wearable devices.

ingestible sensors assist clinicians in identifying colon cancer and irritable bowel syndrome (IBS) [19–22].

Machine-to-machine connectivity, data transfer, sharing, and interoperability enable IoT healthcare. IoT devices report and analyse real-time data, reducing storage. Doctors use medical IoT devices to monitor vital signs in real time, while smartphone apps and intelligent sensors alert patients to essential symptoms [23–24]. Chronic illnesses are rising rapidly, raising healthcare costs. A doctor using a cloud platform can get real-time health data from smartphone-connected medical equipment. A mobile healthcare system, unique IoT technology, and future healthcare features automate patient care [25]. IoTs notify, monitor, and track in real-time, providing hands-on treatments, accuracy, faster medical intervention, and enhanced patient care. IoTs help collect bulk data, which would take years with manual collection [26].

Modern healthcare IoT solutions meet several needs. In regular healthcare facilities, it's tough to manage drugs and ensure patients take them on time. Health IoT speeds up application processors and improves medical picture rendering and delivery [27]. Healthcare monitors patients via wearable IoT devices. Remote IoT patient monitoring collects primary health data from non-hospitalised patients through wearable devices [28].

Advantages

- Personalisation: Doctors can create programs based on patients' needs using software.
- Early diagnosis: Wearable devices allow early detection of symptoms.
- Remote patient monitoring: Healthcare professionals can monitor patients remotely and in real-time using wearable IoT devices.
- Adherence to medication: Wearable devices, with the help of IoT, help patients take their medicines on time and inform medical professionals if they fail.
- Information registry: Real-time data storage allows for exhaustive analysis of patients' medical history.
- Optimal decision-making: Doctors can compare and analyse data for better clinical decisions.
- Cost savings: Remote patient monitoring and medical wearables use IoT technologies to decrease cost, increase accuracy, and ensure regulatory compliance.

8.3.3 SMARTPHONE APPLICATIONS

Smartphone applications have greatly revolutionised healthcare, empowering people to participate actively in the management of their health. The main characteristics comprise health monitoring and tracking, tailored health suggestions, virtual consultations, secure messaging, educational materials, motivation, and goal establishment, as well as self-management of chronic illnesses.

Smartphone apps make maintaining personal and professional relationships and activities practical, dynamic, and engaging. Intelligent apps employ AI components, including data analytics, Machine Learning, deep learning, Natural Language Processing, robotics, general intelligence, expert systems, etc., to provide a new approach to customised health. It helps determine the best cause of symptoms [29]. Intelligent applications are entering economics, technology, media, lifestyle, and healthcare to assist us in doing precise jobs.

Smartphone applications have a profound effect on enhancing patient empowerment. However, there is a need for ongoing enhancements in areas such as accessibility, user-friendliness, and data security. The healthcare information management mobile app made by Doukas et al. is an example of a cloud-based m-health app [30]. Its job is to store, organise, and get Electronic Health Records (EHRs), as well as Digital Imaging and COmmunications in Medicine (DICOM) and Joint Photographic Experts Group (JPEG) medical pictures, in the cloud. It was still just a test, though, so security wasn't a big deal. The cloud-based structure that Elazhary [31] suggests for innovative, context-aware mobile user experiences in healthcare apps is another example. As was already said, these kinds of platforms change what they can do depending on the situation. This is very important in the healthcare field, where people are always going from one place to another, like the office, the patient's room, and the surgery room, at different times and in different weather. In this system, the cloud is used to store user preferences and make assumptions about those preferences so that less experienced staff can use them. The Medical Cloud Multi-Agent

System (MCMAS) was created by Hanen et al. [32] for a polyclinic in Tunisia. It lets mobile users request cloud-based medical services through a mobile app. Patients can use these services like appointments and online consultations, and guardians can use them to see the results of medical analyses and look at medical images. The bots in the cloud handle the exchanges that need to happen.

Benefits

- Increases healthcare access through easy appointment planning, medical information, and telehealth.
- Allows active patient participation in their healthcare journey.
- Provides personalised healthcare by gathering and analysing user data.
- Reduces healthcare costs and boosts efficiency by promoting early intervention.

Challenges

- Data privacy and security: Mobile health apps must be able to protect patient data from illegal access and breaches.
- Health discrepancies: Access to healthcare apps may not be available to certain groups due to technical and internet access gaps.
- User adoption and digital literacy: Apps may be complex for non-technical persons and doctors.
- Regulations and laws: Health apps must meet the laws ensuring high patient safety and data standards.

8.3.4 Artificial Intelligence

By aiding in diagnosis, treatment strategies, and patient assistance, AI is revolutionizing the healthcare. AI-powered chatbots and virtual assistants support patient inquiries, provide medical information, and assist in medicine delivery. AI is also enhancing patient care and security by generating understandings and improving decision-making [33, 34]. As AI technology advances, its applications in analysing medical images, X-rays, scans, diagnosing medical conditions, and formulating treatment plans are expanding. Healthcare administrators worldwide face challenges such as increasing demand for services, improving patient care quality, building more adaptable institutions, and controlling costs. AI uses Machine Learning, NLP, and Deep Learning to analyse massive data sets, reveal health trends, and improve therapy. It can diagnose diseases early, speed up patient healing, and evaluate medical history and scans.

AI chatbots are essential for patient engagement in healthcare, providing triage and symptom checks, medication reminders, individualised health education, behavioural health support, appointment scheduling, data collection and analysis, accessibility, and healthcare professional assistance. They improve chronic disease management with fast answers, individualised treatment, and data analysis. AI systems can learn from patient interactions, adjust talks and health recommendations, and evaluate patient-reported data [35–38].

Benefits

- Increased efficiency and productivity: Task automation enhances workflow and optimises resource use.
- Better diagnosis and treatment: AI algorithms provide precise diagnosis and individualised treatment planning, providing prompt interventions and optimal treatment alternatives.
- Cost reduction and resource management: AI optimises resource allocation, finds inefficiencies, and forecasts maintenance needs, sustaining healthcare.
- Advanced-data analysis and insights: AI can find trends and patterns in complicated datasets for individualised therapy and evidence-based medicine.
- Better patient experience and engagement: AI-powered virtual assistants offer 24/7 support and real-time health tracking.

Key Challenges

- Data privacy and security: Ensuring compliance with legislation, preventing unwanted access, and managing possible breaches.
- Integration and Interoperability: Standardisation and data-sharing standards enable smooth integration and cooperation.
- Trust and ethical activity: Compliance with legislation and moral principles is essential.
- Insufficient and poor data: Data collection, storage, and biases might impede AI applications.
- Workflow integration and adoption: Healthcare workers must be trained on AI use, overcome change resistance, and manage job displacement.

8.3.5 EDGE COMPUTING

In edge computing environment, networking, storage, and computation are brought closer to where data is generated and consumed. This proximity proves beneficial in critical care and life-threatening situations. With edge computing, physicians can remotely monitor patients' vital signs and devise action plans promptly. The challenge of delivering health treatment in remote rural areas persists despite advancements in telemedicine and the availability of health data. Doctors have struggled to offer timely, quality care to patients in these areas with poor internet connections. IoT edge devices generate substantial amounts of data, which is advantageous, but it also poses challenges for healthcare practitioners in terms of managing and safeguarding it. The data often lacks organisation and clarity, overwhelming cloud infrastructures. Consequently, these infrastructures become ill-prepared to execute sophisticated analytics algorithms for organising and extracting valuable information [39, 40]. Figure 8.3 shows the architecture of edge computing.

Healthcare IT infrastructures can obtain health data and access real-time information to diagnose and address health issues by maintaining critical processing activities on network edge devices. IoT devices play a pivotal role in observing an

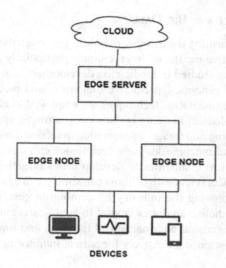

FIGURE 8.3   Edge computing architecture.

individual's current health status and alerting them to irregularities. This capability allows for faster response times, potentially saving lives. In the realm of healthcare IT infrastructures, the significance of edge computing enterprises cannot be overstated. These gadgets contribute significantly to enhancing patient experiences by facilitating improved connectivity and accessibility to essential health information [41].

Many hospitals provide streaming movies, games, and interactive educational programs. Edge data centres can help decentralise and distribute this content with minimal latency. Edge computing will make data organisation and classification easier and secure data sharing for colleagues, letting academics use hitherto untapped data to develop technology. Edge computing is helpful for devices that need to act on data immediately but cannot be sent to the cloud [42].

Benefits

- Processing data closer to the network's edge enhances speed, efficiency, and reliability.
- Supports telemedicine and real-time health monitoring.
- Improves data privacy and ensures compliance with legislation.
- Reduces the need for long-distance data transmission, enhancing patient privacy.
- Reduces network stress and bandwidth usage, ensuring the reliability of remote health applications.
- Scalable and cost-effective for deploying health monitoring systems without overwhelming central servers.
- Supports AI and machine learning for providing individualised health advice to patients on their devices.
- It is expected to significantly shape patient-centred care paradigms and enhance patient involvement in healthcare processes as the industry digitises.

8.3.6 DATA SCIENCE AND BIG DATA

Data science is transforming the healthcare industry through the provision of precise diagnostics and effective medicinal interventions, particularly in the field of medical imaging. It is being studied in medication development, novel drug development and delivery, testing, genomics, predictive diagnosis, and medical imaging. Deep learning and image segmentation techniques are employed to identify abnormalities in scanned images. Medical imaging is the most promising application of data science, with deep learning and image segmentation detecting flaws in scanned images [43]. Data science and machine learning techniques enhance prediction accuracy, enabling pharmaceutical companies to develop drugs targeting genomic sequence primary mutations. Data science algorithms can simulate drug effects, avoiding long experiments and improving the industry by combining genetics and drug-protein binding datasets. Predictive analytics is vital in healthcare, enhancing patient care, chronic disease management, pharmaceutical logistics, and supply chain efficiency. IoT devices use data science extensively for patient monitoring and disease prevention [44, 45].

Economic healthcare spending is optimised via data science. Analytical devices allow doctors to track a patient's vital signs and calorie consumption. They use home and wearable sensors to monitor patients' health. Many systems for chronically ill individuals track their movements, check their physical parameters, and analyse data trends. It predicts patient issues based on current health using real-time data. It also helps doctors make decisions to aid distressed patients and predict disease progression. Data science improves healthcare workflow and diagnostic and treatment productivity and has affected the healthcare industry. Data science analytical methods learn from past data and predict outcomes accurately. Virtual assistant apps demonstrate data science well. Data scientists have built customised patient experience platforms. Medical applications using data science assist in identifying illnesses by assessing symptoms. Just enter symptoms, and the algorithm will predict the patient's diagnosis and condition. Based on the patient's circumstances, it can prescribe precautions, drugs, and therapy [46].

Altogether, data science turns massive health data into meaningful insights, improving patient involvement. Personalised patient care is enabled by evaluating patient data, anticipating health risks and consequences, and improving health outcomes. It can also find successful illness management techniques based on patient data. Data science enables individualised health communications and instructional material, enhancing communication and education. By anticipating patient volume, scheduling appointments, and lowering wait times, it optimises healthcare services. Remote monitoring and telehealth benefit from real-time data analysis and appointment optimisation. Sentiment analysis and experience customisation assist in comprehending patient comments and experiences. Data science helps manage population health by examining patterns and directing community health activities. Data science in patient interaction is predicted to rise as technology progresses, creating new healthcare innovations and improvement opportunities.

Nearly every industry uses big data, but healthcare benefits most. Big data helps analyse medical treatments from pre-to post-diagnosis. A solution is found by comparing causes, symptoms, and therapies. Big data finds effective and standardised illness treatments. Correlating pharmacological side effects, categorising symptoms for diagnosis, and designing effective treatment chemicals and combinations are also involved. Big data manages healthcare expenditures, improves illness management, and boosts results. It helps healthcare management improve diagnosis and treatment. Big data allows healthcare organisations to analyse practice patterns, resource use, and hospital costs. Big data stores large volumes of data systematically. Medical professionals make decisions with so much data [47].

Big data streamlines healthcare administration. It reduces care measurement costs, provides the finest professional help, and guides at-risk patients. It helps doctors analyse data from multiple sources. Big data helps health administrators avoid human mistakes like wrong dosages, drugs, and more. Insurance companies also benefit greatly. They stop many insurance fraud claims. Clinical data is hospital-generated big data such as medical imaging and clinic data. Big Data in medicine from fundamental clinical operations significantly affects the medical industry [47, 48].

Benefits

- Big Data helps healthcare practitioners make better decisions by analysing patient data in real time.
- It enables personalised medicine by customising treatments based on genetic data, medical history, and lifestyle variables.
- Predictive analytics can forecast health trends, identify high-risk patients, and improve healthcare efficiency.
- Big Data accelerates medical research by aggregating data for drug discovery and clinical trials.
- Continuous monitoring and remote care are possible through wearables and remote monitors.
- Big Data solutions save costs by streamlining operations and reducing unnecessary testing.
- Public health initiatives benefit from Big Data by tracking infectious diseases and health inequities.
- Patients can engage better with their health through personalised advice and interactive apps.

Challenges

- Protect patient data from breaches and unauthorised access.
- Ensure smooth data sharing and integration for effective big data analytics.
- Regularly validate and clean data for accuracy in decision-making.
- Overcome economic obstacles for smaller healthcare organisations to implement big data solutions.

- Address ethical and regulatory challenges in data sharing and storage laws.
- Improved data security and privacy are essential due to the increasing volume of healthcare data.

8.3.7 BLOCKCHAIN

The healthcare business is being transformed by blockchain technology, which allows safe and interoperable data exchange and provenance tracking and reduces delays, costs, and mistakes. It can connect complicated healthcare, finance, and payment systems, monitor drugs from manufacturer to patient, combat counterfeiting, and increase supply chain medication traceability. Blockchain applications encrypt data in distributed ledger blocks. They aid pharmaceutical firms with supply chain, clinical trials, and IoT device compatibility management. Blockchain ledgers help authorities discover counterfeit medications, while blockchain transactions track materials and compliance. Blockchain technology allows authorised parties to access patient data via a decentralised network, saving time and money on translation and reconciliation. Blockchain networks secure patient privacy and speed up medical qualification certification. This cooperation might transform healthcare worldwide by tackling security and transparency issues while saving money [49].

Blockchain technology can improve healthcare patient engagement and communication, security, privacy, interoperability, patient control, communication, research, administrative burden reduction, trust enhancement, cost reduction, population health improvement, and personalised medicine support. Regulatory and legal frameworks, interoperability, data quality and integrity, scalability, privacy issues, and patient adoption are obstacles to adoption. Blockchain technology is being tested in several industries, including healthcare [50, 51], where it might transform patient data management, sharing, and communication. Blockchain technology makes transaction records safe, transparent, and tamperproof. Blockchain can increase patient involvement and communication by boosting data security and privacy, healthcare provider interoperability, and patient sovereignty over their health data [52].

Benefits

- Enables security and privacy by creating unchangeable blocks of information.
- Facilitates interoperability through intelligent contracts and unified data standards.
- Enables data accuracy through recording and time stamping of every activity.
- Enables better patient data sharing, ensuring safe sharing.
- Automates routine tasks, enhancing healthcare worker efficiency.
- Facilitates drug traceability, reducing the risk of fake drugs.
- Facilitates regulatory compliance, enhancing transparency in the pharmaceutical business.

Challenges

- Integration issues: Integration of new systems with existing ones can be challenging and costly.
- Scaling issues: Blockchain networks may slow down transactions and reduce efficiency.
- Quality of data: Import to have high-quality data for data purity.
- Need for user education and adoption: Healthcare workers and patients may need training on blockchain-based systems.
- Safety concerns: 51% of smart contracts have security flaws [53].

8.4   REAL-WORLD CASE STUDIES: A FEW INSTANCES

CASE STUDY 1: PROJECT ECHO (EXTENSION FOR COMMUNITY HEALTHCARE OUTCOMES)

BACKGROUND

It was established in 2003 by Dr. Sanjeev Arora at the University of New Mexico. It aims to make the availability of specialist healthcare services in rural and neglected areas. It is a lifelong learning and guided practice model that changes the way medical education works and dramatically expands the ability of the workforce to provide best-practice specialty care and lower health disparities. The hub-and-spoke knowledge-sharing networks are what the ECHO model is all about. They are run by expert teams who use multipoint videoconferencing to hold virtual sessions with community doctors. This helps doctors, nurses, and other healthcare workers in primary health sectors, to learn on giving excellent care to patients in their places [54–56].

The ECHO Model: Project ECHO is a network of virtual clinics that cluster on particular diseases or ailments.

- Case-based learning: In case-based learning sessions, primary care providers (PCPs) from participating communities team up and solve real-world patient problems through expert discussions and collaborative learning. This enhances their local knowledge and expertise on the subject and fosters effective problem-solving capabilities.

IMPACT OF PROJECT ECHO

Enhanced patient outcomes: Research indicates that Project ECHO improves patient outcomes by:

- Enhancing primary care provider knowledge and abilities in handling complex conditions

- Reducing healthcare disparities by providing specialist expertise in underserved areas
- Decreasing hospital admissions and readmissions for individuals effectively managing chronic diseases

Project ECHO achieves these benefits through telehealth technology, which:

- Makes it more convenient for people to access high-quality medical care by reducing travel distances to see an expert.
- Offers a cost-effective solution to enhance healthcare in resource-limited settings [54–56].

Global Impact: Project ECHO has been introduced in several nations worldwide, targeting a wide range of healthcare needs. It is a powerful tool for encouraging healthcare workers and improving access to quality treatment in various settings, thanks to its scalability and versatility. Project ECHO has expanded its focus to cover different areas such as Hepatitis C, mental health issues, and substance abuse [56].

Overall, Project ECHO is a highly effective model for improving healthcare services in underserved areas. This initiative boosts the authority and impact of primary care doctors, improves patient outcomes, and helps reduce healthcare disparities through the use of technology and collaborative learning [57].

CASE STUDY 2: IBM WATSON HEALTH-
AI-DRIVEN HEALTHCARE

BACKGROUND

IBM Watson Health [58], an artificial intelligence (AI) and big data analytics tool, analyses large amounts of structured and unstructured healthcare data, including:

- Electronic health records (EHRs)
- Clinical trial data
- Medical literature and research studies
- Genomic data

By analysing these distinct datasets, Watson Health proposes the improvement to various aspects of healthcare, including:

- **Diagnosis and treatment:** The platform helps identify potential diagnoses and suggest personalised treatment options based on individual patient data and evidence-based insights.

- **Clinical research and drug discovery:** The development of personalised medicinal approaches, analysis of huge datasets for researchers, promotion of drug discovery are all supported by IBM Watson Health.
- **Population health management:** The platform allows healthcare organisations to identify at-risk individuals and communities, enabling proactive interventions and preventive care measures.

IBM Watson Health supports healthcare providers with comprehensive insights and evidence-based recommendations. The platform facilitates personalised treatment plans that consider individual patients' unique medical history and genetic makeup. AI-driven tools help in improving healthcare workflows, automating tasks and thereby reducing administrative burdens. Patients get the advantage of having increased access to essential information through the portals or educational resources [59–61].

Challenges

- **Ethical considerations:** As always there is an ethical concern in using AI in healthcare. For this reason, IBM Watson Health tries to be open and highlights its algorithmic objectivity throughout the development and deployment of its AI models [62].
- **Data privacy and security:** Ensuring the security and privacy of sensitive patient data is crucial. IBM Watson Health complies with relevant data privacy regulations and follows strict data security protocols.
- **Limited data access and interoperability:** Still there are challenges in data access and interoperability among different healthcare systems limiting the effectiveness of AI tools.

FUTURE OF IBM WATSON HEALTH

IBM Watson Health is designing new AI applications by collaborating with stakeholders across the healthcare sectors. It is vital to have responsible and ethical considerations in ensuring patient-centric and equitable utilisation of AI technology.

CASE STUDY 3: MEDILEDGER-TRANSFORMING HEALTHCARE DATA MANAGEMENT WITH BLOCKCHAIN

BACKGROUND

MediLedger [63] is one blockchain network with restrictive access and aims to establish the first decentralised network for the pharmaceutical industry.

The MediLedger network leverages blockchain technology in developing and executing cross-industry business operations and to verify communications exchanged among the participants. The network leverages a set of standardized and interoperable protocol primitives to make the data exchange smooth. Users share information with only those partners of their choice, using permission-based private messaging. The interactions with the leading pharmaceutical industry partners are also supported. It also provides solutions to secure management of healthcare data.

KEY FEATURES

- **Secure data storage:** For developing a safe and tamper-proof platform for storing and sharing patient medical records, MediLedger is using blockchain technology. It ensures data integrity and reduces the risk of unauthorized access, changes or data breaches.
- **Patient data ownership:** MediLedger provides features for patients to control their healthcare data. Patients have the right to grant access of their records, to selected healthcare professionals or institutions, thus fostering patient autonomy and transparency in the use of data.
- **Streamlined data exchange:** A secure and efficient data exchange between healthcare providers is facilitated. That is, with permission, authorised individuals from other healthcare institutions can access the patient records. The level of collaboration and care coordination can be thus improved.
- **Improved medication management:** Tracking the medication adherence and adverse reactions facility ensures enhanced medication management and drug safety monitoring.

POTENTIAL BENEFITS

- **Improved data security:** Blockchain technology offers a high level of security, reducing the risk of data breaches and unauthorized access to sensitive patient information.
- **Increased patient engagement:** The fact that patients are able to control their data makes them more engaged in the medical decisions of their lives and induces trust in the health providers.
- **Better care coordination:** Ease of data sharing allows for collaborative care with more informed, efficient decisions.
- **Reduced administrative burden:** Secure, efficient data exchange can perhaps decrease administrative burden and optimize healthcare workflow for healthcare providers.

CHALLENGES

- **Low adoption:** Blockchain technology in healthcare is in its evolutionary stage, and a higher level of adoption can be achieved by solving problems related to scalability, interoperability, and interfacing with existing healthcare IT systems.
- **Regulatory landscape:** Data privacy and security regulations have changed, and the same will be taken into consideration when the blockchain is introduced in healthcare.
- **Cost and infrastructure:** Implementation of blockchain infrastructure is expensive, and industry-wide standards would have to be built for large-scale adoption.

FUTURE OF MEDILEDGER

MediLedger is continuously evolving and is currently looking for new use cases beyond the medication supply chain. Blockchain technology can make data more secure, give patients more power, and thereby improve overall healthcare in a digital environment.

CASE STUDY 4: ALIVECOR KARDIAMOBILE-ADVANCING HEALTHCARE USING AI AND MACHINE LEARNING

BACKGROUND

The AliveCor KardiaMobile, is a smartphone-connected ECG gadget that enables users to record their electrocardiogram (ECG) and share it with their doctor for prompt analysis. This allows people to assume control of their heart health and promptly seek medical care if necessary. KardiaMobile is a handheld device that transforms electrical ECG signals from electrodes situated on a metallic plate to ultrasound signals following a 30-second finger contact. These signals are then transmitted to a smartphone in the form of a single-lead ECG. Critically, the KardiaMobile 6L includes a third electrode that can be positioned on the left ankle or knee in order to simulate the six traditional limb leads of a complete 12-lead ECG. The patient and healthcare provider have access to the encrypted cloud where the recorded data is stored. Research has shown that using KardiaMobile can help in the early diagnosis of atrial fibrillation, a cardiac arrhythmia that can be life-threatening [64–67].

The systematic review of the AliveCor KardiaMobile device for atrial fibrillation (AF) screening found the following [65]:

- Atrial fibrillation monitoring with the AliveCor device is accessible, valid, and practical.
- Study population characteristics affected atrial fibrillation detection rates. Older age, targeted locations, and chronic illness populations had more atrial fibrillation cases.
- Feasibility metrics showed that the AliveCor device improves AF screening procedures, resources, and management.
- Screening length affected atrial fibrillation detection rates.
- The papers evaluated were heterogeneous, highlighting the need for more standardised protocols and techniques in future studies.
- The AliveCor KardiaMobile can be used for opportunistic and systematic atrial fibrillation screening.

These demonstrate KardiaMobile's convenience, validity, and practicality in clinical practice and research for atrial fibrillation screening.

CASE STUDY 5: FITBIT-EMPOWERING INDIVIDUALS THROUGH WEARABLE TECHNOLOGY

BACKGROUND

Fitbit, founded in 2007, is a pioneer in the wearable technology market, offering a range of fitness trackers and smartwatches. These devices track various health metrics, including steps taken, distance travelled, heart rate, sleep patterns, and calorie expenditure. Fitbit aims to empower individuals to take charge of their health and well-being by providing personalised data insights and motivating them to adopt healthier lifestyles [68].

TECHNOLOGY AND FEATURES

- **Activity Tracking:** Fitbit devices track daily steps, distance, active minutes, and calories burned, allowing users to monitor their activity levels and set achievable goals.
- **Sleep Monitoring:** Advanced models track sleep stages (light, deep, and REM) and provide insights into sleep quality, helping users understand their sleep patterns and identify potential sleep disturbances.
- **Heart Rate Monitoring:** Continuous heart rate monitoring allows users to track their heart rate throughout the day, learn about resting heart rate trends, and potentially identify potential heart health concerns.

- **Additional Features:** Depending on the model, Fitbit devices offer additional features like:

  - GPS tracking for outdoor activities
  - Stress level monitoring
  - Guided breathing exercises
  - Personalised health coaching programs
  - Mobile app integration for data visualisation and analysis

IMPACT ON USER EMPOWERMENT

- **Increased self-awareness:** By tracking health metrics, users gain valuable insights into their activity levels, sleep patterns, and overall well-being, fostering self-awareness and motivation for improvement.
- **Goal setting and behaviour change:** Fitbit devices encourage users to set personalised goals, track progress, and celebrate achievements, motivating them to adopt healthier behaviours and become more physically active.
- **Wellness management:** Features like sleep monitoring and stress management tools can help users understand their overall well-being and take proactive steps to improve their mental and physical health.

RESEARCH AND EVIDENCE

Studies [69–71] have shown that Fitbit use can lead to:

- **Increase in physical activity:** Analysis of randomized trials showed that Fitbit use produced a significant increase in the number of steps taken daily compared to control groups.
- **Improved weight control:** Using Fitbit is credited to boost weight loss and maintain weight.
- **Improved sleep:** Research indicates that the use of sleep monitoring features is able to enhance sleep quality.

CHALLENGES AND LIMITATIONS

- **Data privacy:** Fitbit emphasises data security measures and user control over data sharing as data security and privacy is a major concern in collecting personal health information.
- **Accuracy of technology:** Users should be aware of limitations of the data from wearable devices, and should therefore consult healthcare professionals for any medical concerns.
- **Accessibility and affordability:** Not all can afford high-end models. This can lead to to a digital divide in health access.

FUTURE OF FITBIT

As part of improvement, Fitbit is integrating new features like blood oxygen monitoring, advanced sleep analysis etc. It is also trying to track personalised user experiences and expand to diverse health requirements.

8.5 PROS AND CONS OF DIGITAL TOOLS IN PATIENT ENGAGEMENT

There are several advantages that accrue to healthcare with regard to the use of digital technologies: personalized care, better efficiency, better patient experience, access to information, collaboration and communication, better diagnosis and treatment, online education, data management and analysis, secure and transparent data exchange, and cost-saving. These technologies enable healthcare providers to tune their service provision to suit individual patients, yield better results.

However, the adoption of technology also has challenges related to patient engagement. These consist of issues to do with accessibility and equity where not all patients have the technology and/or internet connectivity to use the digital tools, which thus increases the digital gap and further areas of healthcare disparities. Privacy and security issues are also present, whereby issues of data breaches and unauthorised access to personal health information risk the person. Reliability and accuracy are also a challenge, given that sometimes technology may erode patients' engagement and individuals' treatment by differentiating them. Healthcare workers must be trained and supported to use and embrace these technologies, as the speed at which technology advances may make it hard for people to catch up and use all its capabilities.

Cost is also a significant issue since such technologies are costly in development, adoption, and maintenance and therefore may deprive some patients of receiving care. However, despite these challenges, the adoption of digital technologies in healthcare can be better for patient engagement, increased efficiency, and better communication among the various stakeholders in healthcare.

8.6 FUTURE TRENDS IN PATIENT ENGAGEMENT WITH TECHNOLOGY

The future of patient engagement is full of technology innovations as healthcare moves toward personalized, proactive care. Artificial Intelligence and Machine Learning make predictive, tailored care as well as AI-driven communication tools. Wearables, IoT devices—will transform health monitoring by pulling in data from wearables into electronic health records. Telehealth and virtual care will continue to grow, rendering online visits more immersive. Health data storage and sharing will be reformed using blockchain technology, bringing security and transparency. Personalized medicine and genomics will integrate into routine care, to get the right treatment for the right patient. Evidence-based interventions will be delivered using digital therapeutics and mobile health apps. There will be enhanced self-management tools as patients are empowered through education and communities by interactive platforms and online patient communities. Smart homes and ambient assisted living technologies ensure independent living. The development of regulatory and ethical frameworks will be needed to ensure that the new technologies used in healthcare are safe and ethical.

8.7 CONCLUSION

Indeed, digital tools lead to patient engagement in healthcare. The most important tools include data visualization tools, real-time treatment solutions, AI-enabled chatbots for bookings and bill payments, wearable devices for health monitoring and preventing chronic diseases, internet-based healthcare IT systems for better delivery of care and education, blockchain technology for secure storage and interchange of health data, data science, and big data. These tools create a transparent view of health data, offer immediate care, and simplify administrative work. They also assist in communication efficiency, building medical information management trust, and cost saving. As the healthcare landscape evolves, these digital tools will play a very significant role in shaping a future where patient engagement and empowerment will be at the forefront of care delivery.

The other healthcare relevant technological advancements are Fitbit, Project ECHO, IBM Watson Health, and AliveCor KardiaMobile, among others. They have improved access to healthcare, patient outcomes, and overall patient health. Project ECHO improves the competencies of primary care providers through virtual clinics. IBM Watson Health, as well as drug discovery, is supported by tailored treatment plans. MediLedger uses blockchain technology in managing data. AliveCor KardiaMobile uses smartphone-based remote electrocardiogram monitoring. Fitbit generates self-awareness by using an activity tracker. However, it has the aspect of accessibility, data privacy, and ethical issues. The advantages of blockchain technology are increased data sharing, chain of control tracking, unified data standards, data integrity, and regulatory compliance. The issues with smart contracts in blockchain are regulatory compliance, scalability, data quality, user education, and security vulnerabilities.

REFERENCES

[1] G. Kaur, M. Gupta, and R. Kumar, "IoT based smart healthcare monitoring system: A systematic review," *Annals of the Romanian Society for Cell Biology*, vol. 25, pp. 3721–3728, Feb. 2021, Accessed: May 11, 2024. [Online]. Available: http://annalsofrscb.ro/index.php/journal/article/view/484.

[2] "Computational Intelligence in Healthcare: Applications, Challenges, and . . . —Google Books," Accessed: March 11, 2024. [Online]. Available: https://books.google.co.in/books?hl=en&lr=&id=O5yaEAAAQBAJ&oi=fnd&pg=PP1&dq=info:nWvoZzMvTdQJ:scholar.google.com&ots=ip6IAhJwuL&sig=uQHG8X3alHG2cRznIJPgKrfk5T8&redir_esc=y#v=onepage&q&f=false.

[3] D. Lewin, and S. Piper, "Patient empowerment within a coronary care unit: Insights for health professionals drawn from a patient satisfaction survey," *Intensive and Critical Care Nursing*, vol. 23, no. 2, pp. 81–90, Apr. 2007, doi: 10.1016/j.iccn.2006.09.003.

[4] L. P. Fumagalli, G. Radaelli, E. Lettieri, P. Bertele, and C. Masella, "Patient empowerment and its neighbours: Clarifying the boundaries and their mutual relationships," *Health Policy*, vol. 119, no. 3, pp. 384–394, Mar. 2015, doi: 10.1016/j.healthpol.2014.10.017.

[5] M. M. Funnell *et al.*, "Empowerment: An idea whose time has come in diabetes education," *The Diabetes Educator*, vol. 17, no. 1, pp. 37–41, Feb. 1991, doi: 10.1177/014572179101700108.

[6] Y. Chen, and I.-C. Li, "Effectiveness of interventions using empowerment concept for patients with chronic disease: A systematic review," *JBI Library of Systematic Reviews*, vol. 7, no. 27, pp. 1179–1233, Jan. 2009, doi: 10.11124/jbisrir-2009–208.

[7] P. Bravo, A. Edwards, P. Barr, I. Scholl, G. Elwyn, and M. McAllister, "Conceptualising patient empowerment: A mixed methods study," *BMC Health Services Research*, vol. 15, no. 1, Jul. 2015, doi: 10.1186/s12913–015–0907-z.

[8] G. Iyawa, M. Herselman, and A. Botha, "Digital health innovation ecosystems: From systematic literature review to conceptual framework," *Procedia Computer Science*, vol. 100, pp. 244–252, Jan. 2016, doi: 10.1016/j.procs.2016.09.149.

[9] P. Y. K. Chau, and P. J.-H. Hu, "Investigating healthcare professionals' decisions to accept telemedicine technology: An empirical test of competing theories," *Information & Management*, vol. 39, no. 4, pp. 297–311, Jan. 2002, doi: 10.1016/s0378–7206(01)00098–2.

[10] B. Moazzami, N. Razavi-Khorasani, A. D. Moghadam, E. Farokhi, and N. Rezaei, "COVID-19 and telemedicine: Immediate action required for maintaining healthcare providers well-being," *Journal of Clinical Virology*, vol. 126, p. 104345, May 2020, doi: 10.1016/j.jcv.2020.104345.

[11] R. Chunara *et al.*, "Telemedicine and healthcare disparities: A cohort study in a large healthcare system in New York City during COVID-19," *Journal of the American Medical Informatics Association*, vol. 28, no. 1, pp. 33–41, Aug. 2020, doi: 10.1093/jamia/ocaa217.

[12] H. Leite, I. R. Hodgkinson, and T. Gruber, "New development: 'Healing at a distance'— telemedicine and COVID-19," *Public Money & Management*, vol. 40, no. 6, pp. 483–485, Apr. 2020, doi: 10.1080/09540962.2020.1748855.

[13] S. R. Chowdhury, T. C. Sunna, and S. Ahmed, "Telemedicine is an important aspect of healthcare services amid COVID-19 outbreak: Its barriers in Bangladesh and strategies to overcome," *The International Journal of Health Planning and Management*, vol. 36, no. 1, pp. 4–12, Aug. 2020, doi: 10.1002/hpm.3064.

[14] M. S. Jalali, A. B. Landman, and W. J. Gordon, "Telemedicine, privacy, and information security in the age of COVID-19," *Journal of the American Medical Informatics Association*, vol. 28, no. 3, pp. 671–672, Dec. 2020, doi: 10.1093/jamia/ocaa310.

[15] M. J. I. Mainguyague *et al.*, "Telemedicine for postoperative follow-up, virtual surgical clinics during COVID-19 pandemic," *Surgical Endoscopy and Other Interventional Techniques*, vol. 35, no. 11, pp. 6300–6306, Nov. 2020, doi: 10.1007/s00464-020-08130-1.

[16] C. White-Williams, and D. Shang, "Telehealth for chronic disease management among vulnerable populations," *Journal of Racial and Ethnic Health Disparities*, pp. 1–8, 2023. doi: 10.1007/s40615-023-01588-4.

[17] S. Patel, H.-K. Park, P. Bonato, L. Chan, and M. M. Rodgers, "A review of wearable sensors and systems with application in rehabilitation," *Journal of Neuroengineering and Rehabilitation*, vol. 9, no. 1, Apr. 2012, doi: 10.1186/1743-0003-9-21.

[18] J. Bandodkar, I. Jeerapan, and J. Wang, "Wearable chemical sensors: Present challenges and future prospects," *ACS Sensors*, vol. 1, no. 5, pp. 464–482, May 2016, doi: 10.1021/acssensors.6b00250.

[19] H. C. Ates *et al.*, "End-to-end design of wearable sensors," *Nature Reviews Materials,* vol. 7, no. 11, pp. 887–907, Jul. 2022, doi: 10.1038/s41578–022–00460-x.

[20] Y. Cheng, K. Wang, H. Xu, T. Li, Q. Jin, and D. Cui, "Recent developments in sensors for wearable device applications," *Analytical and Bioanalytical Chemistry,* vol. 413, no. 24, pp. 6037–6057, Aug. 2021, doi: 10.1007/s00216-021-03602-2.

[21] T. Lončar-Turukalo, E. Zdravevski, J. M. Da Silva, I. Chouvarda, and V. Trajkovik, "Literature on wearable technology for connected health: Scoping review of research trends, advances, and barriers," *Journal of Medical Internet Research,* vol. 21, no. 9, p. e14017, Sep. 2019, doi: 10.2196/14017.

[22] L. Manjakkal, S. Dervin, and R. Dahiya, "Flexible potentiometric pH sensors for wearable systems," *RSC Advances,* vol. 10, no. 15, pp. 8594–8617, Jan. 2020, doi: 10.1039/d0ra00016g.

[23] S. Park, and S. Jayaraman, "Enhancing the quality of life through wearable technology," *IEEE Engineering in Medicine and Biology Magazine,* vol. 22, no. 3, pp. 41–48, May 2003, doi: 10.1109/memb.2003.1213625.

[24] Y. Yin, Y. Zeng, X. Chen, and Y. Fan, "The internet of things in healthcare: An overview," *Journal of Industrial Information Integration,* vol. 1, pp. 3–13, Mar. 2016, doi: 10.1016/j.jii.2016.03.004.

[25] B. B. Zarpelo, R. S. Miani, C. T. Kawakani, and S. C. De Alvarenga, "A survey of intrusion detection in internet of things," *Journal of Network and Computer Applications,* vol. 84, pp. 25–37, Apr. 2017, doi: 10.1016/j.jnca.2017.02.009.

[26] L. Yehia, A. E. Khedr, and A. Darwish, "Hybrid security techniques for internet of things healthcare applications," *Advances in Internet of Things,* vol. 05, no. 03, pp. 21–25, Jan. 2015, doi: 10.4236/ait.2015.53004.

[27] Y. Yin, Y. Zeng, X. Chen, and Y. Fan, "The internet of things in healthcare: An overview," *Journal of Industrial Information Integration,* vol. 1, pp. 3–13, Mar. 2016, doi: 10.1016/j.jii.2016.03.004.

[28] N. Dey, A. S. Ashour, and C. Bhatt, "Internet of things driven connected healthcare," in *Studies in Big Data,* 2017, pp. 3–12. doi: 10.1007/978-3-319-49736-5_1.

[29] H. Sarker, M. M. Hoque, K. Uddin, and T. Alsanoosy, "Mobile data science and intelligent apps: Concepts, AI-based modeling and research directions," *Mobile Networks and Applications,* vol. 26, no.1, pp. 285–303, Sep.2020, doi: 10.1007/s11036–020–01650-z.

[30] C. Doukas, T. Pliakas, and I. Maglogiannis, "Mobile healthcare information management utilising cloud computing and Android OS," *Annual International Conference of the IEEE Engineering in Medicine and Biology Society,* 2010. doi: 10.1109/iembs.2010.5628061.

[31] H. Elazhary, "A cloud-based framework for context-aware intelligent mobile user interfaces in healthcare applications," *Journal of Medical Imaging and Health Informatics,* vol. 5, no. 8, pp. 1680–1687, doi: 10.1166/jmihi.2015.1620.

[32] J. Hanen, Z. Kechaou, and M. B. Ayed, "An enhanced healthcare system in mobile cloud computing environment," *Vietnam Journal of Computer Science,* vol. 3, no. 4, pp. 267–277, Jul. 2016, doi: 10.1007/s40595–016–0076-y.

[33] K. Denecke, M. Tschanz, T. L. Dorner, and R. May, "Intelligent conversational agents in healthcare: Hype or hope?," *PubMed,* vol. 259, pp. 77–84, Jan. 2019, [Online]. Available: https://pubmed.ncbi.nlm.nih.gov/30923277.

[34] A. Bohr, and K. Memarzadeh, "The rise of artificial intelligence in healthcare applications," in *Elseviere Books,* 2020, pp. 25–60. doi: 10.1016/b978-0-12-818438-7.00002–2.

[35] F. Amisha, P. Malik, M. Pathania, and V. K. Rathaur, "Overview of artificial intelligence in medicine," *Journal of Family Medicine and Primary Care,* vol. 8, no. 7, p. 2328, Jan. 2019, doi: 10.4103/jfmpc.jfmpc_440_19.

[36] T. Q. Sun, and R. Medaglia, "Mapping the challenges of Artificial Intelligence in the public sector: Evidence from public healthcare," *Government Information Quarterly,* vol. 36, no. 2, pp. 368–383, Apr. 2019, doi: 10.1016/j.giq.2018.09.008.

[37] O. Asan, A. E. Bayrak, and A. Choudhury, "Artificial intelligence and human trust in healthcare: Focus on clinicians," *Journal of Medical Internet Research,* vol. 22, no. 6, p. e15154, Jun. 2020, doi: 10.2196/15154.

[38] D. Lee, and S. N. Yoon, "Application of artificial intelligence-based technologies in the healthcare industry: Opportunities and challenges," *International Journal of Environmental Research and Public Health,* vol. 18, no. 1, p. 271, Jan. 2021, doi: 10.3390/ijerph18010271.

[39] H. Sodhro, Z. Luo, A. K. Sangaiah, and S. W. Baik, "Mobile edge computing based QoS optimisation in medical healthcare applications," *International Journal of Information Management,* vol. 45, pp. 308–318, Apr. 2019, doi: 10.1016/j.ijinfomgt.2018.08.004.

[40] M. Chen, W. Li, Y. Hao, Y. Qian, and I. Humar, "Edge cognitive computing based smart healthcare system," *Future Generation Computer Systems,* vol. 86, pp. 403–411, Sep. 2018, doi: 10.1016/j.future.2018.03.054.

[41] Md. Z. Uddin, "A wearable sensor-based activity prediction system to facilitate edge computing in smart healthcare system," *Journal of Parallel and Distributed Computing,* vol. 123, pp. 46–53, Jan. 2019, doi: 10.1016/j.jpdc.2018.08.010.

[42] P. Pace, G. Aloi, R. Gravina, G. Caliciuri, G. Fortino, and A. Liotta, "An edge-based architecture to support efficient applications for healthcare industry 4.0," *IEEE Transactions on Industrial Informatics,* vol. 15, no. 1, pp. 481–489, Jan. 2019, doi: 10.1109/tii.2018.2843169.

[43] S. Schulz, R. A. Stegwee, and C. Chronaki, "Standards in healthcare data," in *Springer eBooks,* 2018, pp. 19–36. doi: 10.1007/978-3-319-99713-1_3.

[44] F. Provost, and T. Fawcett, "Data science and its relationship to big data and data-driven decision making," *Big Data,* vol. 1, no. 1, pp. 51–59, Mar. 2013, doi: 10.1089/big.2013.1508.

[45] A. Karim, A. Beni-Hessane, and H. Khaloufi, "Big healthcare data: Preserving security and privacy," *Journal of Big Data,* vol. 5, no. 1, Jan. 2018, doi: 10.1186/s40537-017-0110-7.

[46] K. Kaur, and R. Rani, "Managing data in healthcare information systems: Many models, one solution," *IEEE Computer,* vol. 48, no. 3, pp. 52–59, Mar. 2015, doi: 10.1109/mc.2015.77.

[47] N. Mehta, and A. Pandit, "Concurrence of big data analytics and healthcare: A systematic review," *International Journal of Medical Informatics,* vol. 114, pp. 57–65, Jun. 2018, doi: 10.1016/j.ijmedinf.2018.03.013.

[48] S. Shilo, H. Rossman, and E. Segal, "Axes of a revolution: Challenges and promises of big data in healthcare," *Nature Medicine,* vol. 26, no. 1, pp. 29–38, Jan. 2020, doi: 10.1038/s41591-019-0727-5.

[49] S. Tanwar, K. Parekh, and R. Evans, "Blockchain-based electronic healthcare record system for healthcare 4.0 applications," *Journal of Information Security and Applications,* vol. 50, p. 102407, Feb. 2020, doi: 10.1016/j.jisa.2019.102407.

[50] H. Lei, E.-Y. Choi, and D. Kim, "A novel EMR integrity management based on a medical blockchain platform in hospital," *Electronics,* vol. 8, no. 4, p. 467, Apr. 2019, doi: 10.3390/electronics8040467.

[51] C. Kombe, M. Ally, and A. Sam, "A review on healthcare information systems and consensus protocols in blockchain technology," *International Journal of Advanced Technology and Engineering Exploration,* vol. 5, no. 49, pp. 473–483, Dec. 2018, doi: 10.19101/ijatee.2018.547023.

[52] C. C. Agbo, Q. H. Mahmoud, and J. Eklund, "Blockchain technology in healthcare: A systematic review," *Healthcare,* vol. 7, no. 2, p. 56, Apr. 2019, doi: 10.3390/healthcare 7020056.

[53] CertiK, "Smart contract security: Protecting digital assets—CERTIK—medium," *Medium,* Accessed: Mar. 11, 2023. [Online]. Available: https://certik.medium.com/ smart-contract-security-protecting-digital-assets-719da8a6c646#.

[54] D. Furlan, K. Pajer, W. Gardner, and B. MacLeod, "Project ECHO: Building capacity to manage complex conditions in rural, remote and underserved areas," *Canadian Journal of Rural Medicine,* vol. 24, no. 4, p. 115, Jan. 2019, doi: 10.4103/cjrm.cjrm_20_18.

[55] M. Komaromy, V. Ceballos, A. Zurawski, T. Bodenheimer, D. H. Thom, and S. Arora, "Extension for Community Healthcare Outcomes (ECHO): A new model for community health worker training and support," *Journal of Public Health Policy,* vol. 39, no. 2, pp. 203–216, Dec. 2017, doi: 10.1057/s41271-017-0114-8.

[56] D. Anderson *et al.*, "Improving pain care with project ECHO in community health centers," *Pain Medicine,* vol. 18, no. 10, pp. 1882–1889, Aug. 2017, doi: 10.1093/pm/pnx187.

[57] J. Agley, J. Delong, A. D. Janota, A. Carson, J. S. Roberts, and G. Maupomé, "Reflections on project ECHO: Qualitative findings from five different ECHO programs," *Medical Education Online,* vol. 26, no. 1, Jan. 2021, doi: 10.1080/10872981.2021.1936435.

[58] A. Kumar *et al.*, "A survey on IBM Watson and its services," *Journal of Physics: Conference Series,* vol. 2273, no. 1, p. 012022, May 2022, doi: 10.1088/1742–6596/2273/1/012022.

[59] R. Hoyt, D. Snider, C. J. Thompson, and S. Mantravadi, "IBM Watson analytics: Automating visualisation, descriptive, and predictive statistics," *JMIR Public Health and Surveillance,* vol. 2, no. 2, p. e157, Oct. 2016, doi: 10.2196/publichealth.5810.

[60] "IBM Watson Health," Accessed: Feb. 04, 2024. *IBM Watson Health,* Available: www. ibm.com/.

[61] T. R. Spena, C. Mele, and M. Marzullo, "Practising value innovation through artificial intelligence: The IBM Watson case," *Journal of Creating Value,* vol. 5, no. 1, pp. 11–24, Nov. 2018, doi: 10.1177/2394964318805839.

[62] M. Van Hartskamp, S. Consoli, W. Verhaegh, M. Petković, and A. Van De Stolpe, "Artificial intelligence in clinical health care applications: Viewpoint," *Interactive Journal of Medical Research,* vol. 8, no. 2, p. e12100, Apr. 2019, doi: 10.2196/12100.

[63] "The MediLedger network." Accessed: Feb. 05, 2024. Available: www.mediledger. com/.

[64] A. N. L. Hermans *et al.*, "Mobile health solutions for atrial fibrillation detection and management: A systematic review," *Clinical Research in Cardiology,* vol. 111, no. 5, pp. 479–491, Sep. 2021, doi: 10.1007/s00392-021-01941-9.

[65] V. Savickas *et al.*, "Screening for atrial fibrillation in care homes using pulse palpation and the AliveCor Kardia Mobile® device: A comparative cross-sectional pilot study," *International Journal of Clinical Pharmacy,* Dec. 2023, doi: 10.1007/ s11096–023–01672-z.

[66] I. Goldenthal *et al.*, "Recurrent atrial fibrillation/flutter detection after ablation or cardioversion using the AliveCor KardiaMobile device: iHEART results," *Journal of Cardiovascular Electrophysiology,* vol. 30, no. 11, pp. 2220–2228, Sep. 2019, doi: 10.1111/jce.14160.

[67] Z. P. Girvin, E. S. Silver, and L. Liberman, "Comparison of AliveCor KardiaMobile six-lead ECG with standard ECG in pediatric patients," *Pediatric Cardiology,* vol. 44, no. 3, pp. 689–694, Sep. 2022, doi: 10.1007/s00246-022-02998-7.

[68] "An engaging health and wellness solution for your population," *Fitbit Health Solutions,* Accessed: Feb. 14, 2024. Available: https://healthsolutions.fitbit.com/

[69] C. Cummings, R. Crochiere, A. H. Lansing, R. Patel, and C. Stanger, "A digital health program targeting physical activity among adolescents with overweight or obesity: Open trial," *JMIR Pediatrics and Parenting,* vol. 5, no. 1, p. e32420, Mar. 2022, doi: 10.2196/32420.

[70] A. H. Y. Chu *et al.*, "Comparison of wrist-worn Fitbit Flex and waist-worn ActiGraph for measuring steps in free-living adults," *PLoS One,* vol. 12, no. 2, p. e0172535, Feb. 2017, doi: 10.1371/journal.pone.0172535.

[71] L. M. Feehan *et al.*, "Accuracy of fitbit devices: Systematic review and narrative syntheses of quantitative data," *Jmir Mhealth and Uhealth,* vol. 6, no. 8, p. e10527, Aug. 2018, doi: 10.2196/10527.

9 Exploring Cutting-Edge Technologies Shaping Gender Health

Manoj Kumar Mahto

9.1 INTRODUCTION TO GENDER HEALTH TECHNOLOGIES

In the realm of healthcare, the convergence of technology and science is bringing about significant transformations, particularly with regard to the health of women and men. When we examine the ways in which cutting-edge technologies are transforming the way in which we provide medical care to individuals of varying genders, we enter a world in which digital innovations, medical breakthroughs, and cutting-edge technologies are collaborating to meet the specific health requirements of individuals of all genders. The American Psychological Association (2020) and the World Health Organization (2020) say that "as we learn more about the complicated topic of gender health, it is important to change the way that we talk about it so that we can recognize the variety and diversity that goes beyond simple black and white ideas." In the medical area, the use of technology is changing the way things are done and helping to solve health problems that people of different types have. Peck et al. (2021) say that these improvements not only make it easier for people to get medical care, but they also make the experience more personalized, with the patient being the main focus while everyone is included. Think about methods like DNA tests, telemedicine, and digital health apps. Not only do these things make it easier to get medical care, but they also raise the standard of care and make it easier for everyone to get. In the next section, we will investigate certain aspects of technology and how they influence female health. A further topic that will be discussed is the implications that this has for the future of healthcare (American Medical Association, 2018). Through the process of removing the mystique around these intricate technologies, our objective is to assist individuals in comprehending how the introduction of new technology might result in a more transparent, encouraging, and efficient approach to gender-specific healthcare.

9.1.1 DEFINING GENDER HEALTH AND ITS TECHNOLOGICAL LANDSCAPE

It is essential to have an adequate grasp of the definition of female health technology when diving into the subject at hand. There is a lot more to gender health than just medical phrases. It reflects how a person expresses their gender and how conscious

DOI: 10.1201/9781003473435-9

they are about it. In this bigger picture, we see that men and women have different health needs and situations, and that health is a big part of how someone feels about themselves. Then, the world of gender health technology can be used to create new items that meet these needs. As technology evolves, it will more effectively handle the complicated parts of gender health. Telemedicine systems are being developed to make medical services easier to get as well as apps for digital health provide further details about sexual health and pregnancy. There's more to it than just curing illnesses. There's a whole approach that values and helps everyone on their own unique health path. In this study, we look at how improvements in technology affect many areas of life, including gender health. We want to show how these changes lead to a more open and personalised approach to healthcare for all genders.

9.1.2 IMPORTANCE OF TECHNOLOGICAL ADVANCEMENTS IN GENDER HEALTH

Get a good grasp on the phrase "gender health technologies" before you start to look into them. There's more to gender health than just medical words. It includes how healthy a person is in general when it comes to expressing and identifying their gender. In this bigger picture, we see that men and women have different health needs and situations, and that health is a big part of how someone feels about themselves. Then, the world of gender health technology can be used to create new items that meet these needs. The complex components of gender health have been handled more effectively through technology, which has grown increasingly sophisticated. Telemedicine systems have been developed to make medical care faster to have access to, and mobile health apps provide additional details about sexual health and pregnancy. There's more to it than just curing illnesses. There's a whole approach that values and helps everyone on their own unique health path. In this study, we look at how improvements in technology affect many areas of life, including gender health. We want to show how these changes lead to a more open and personalised approach to healthcare for all genders.

9.2 TELEMEDICINE REVOLUTIONIZING GENDER-SPECIFIC CARE

As the field of healthcare changes all the time, telemedicine has become a big part of that change, especially when it comes to care for different genders. Telemedicine sites aren't just ways to talk to people far away; they're also a big change in how men and women get medical care. Nowadays not do individuals encounter trouble with travel or obtaining a certain thing. In simple terms, it implies that equally men and women can get specialised treatment compared to house. This isn't what this entire shift is about; it's additionally about making treatment easy to get so that everybody, no matter where they live or who they are, can get care that affirms their gender. In women's health, telemedicine is more than just regular visits. It also helps with mental health issues, keeps track of hormone treatment, and makes healthcare more accessible to everyone. Because telemedicine is changing the way women and girls are cared for, it's clear that this change in technology isn't just about the present. It's also about making the future of gender health fairer and easier to get to. Figure 9.1 shows how telemedicine is changing healthcare through virtual care.

FIGURE 9.1   Telemedicine: revolutionizing healthcare through virtual care.

9.2.1   Telehealth Platforms: Enhancing Access and Outreach

Telemedicine services are one of the main things that are changing gender-specific care because they make it easy for more people to get help. These sites don't just let you get medical help; they also make healthcare more open and easier for everyone to get. People of all genders can get specialised care that they have never been able to get before thanks to telehealth. It gets around the problems that come with geography and healthcare gaps. Not only is it helpful that internet talks are easy, but it also gives people the freedom to find gender-affirming care when they need it. Telehealth services do a lot more than just allow people to make occasional trips. On top of that, they can help with mental health problems, stay up to date on hormone treatments, and give you a better knowledge of women's health. There is no doubt that telemedicine has changed care for people of all genders. These tools make healthcare easier to get, more flexible, and more aware of the needs of these people.

9.2.2   Remote Monitoring Devices: Monitoring and Managing Gender Health

Since online tracking gadgets came out, care for different genders has changed. These tools are needed these days to keep an eye on and take care of different parts of gender health. These tools aren't just medical tools; they can do more. People have more control over their health now that they can always keep an eye on and deal with people. Many things about gender health can be found correctly on online tracking systems, such as changes in hormones, monthly cycles, and general health. The method in which treatment is delivered has evolved because of this recent advancement in technology. It offers folks of all genders more control over their medical condition. These devices let you keep track of them all the time, which not only makes particular to gender care more accurate but additionally gives people an over-whelming sense of control and autonomy. Understanding greater detail about how remote monitoring tools affect the well-being of women is an obvious step towards a more customised, prepared, and patient-centred approach to healthcare. This is a big step forward in the continual delivery of based on gender care. It will assist in making persons feel greater autonomy and connected to their personal healthcare. Figure 9.2 shows technological products that help individuals maintain contact and have been easy for everybody to use.

9.3   GENETIC TESTING AND PRECISION MEDICINE IN GENDER HEALTH

When precision healthcare and DNA testing combine their efforts, they can make an enormous impact in the well-being of women and welcome in an entirely novel era of individualised and focused care. As a part of precision medicine, genetic test-ing looks at the unique genetic patterns of both men and women. This can tell you a lot about their sexual health, hormonal balance, and overall health. For this reason, doctors can make treatments that work better and more precisely for each person because they are based on their unique DNA (American Medical Association, 2018). Figure 9.3 shows how genetic medicine and services linked to it are used.

In terms of reproductive health, this new way of thinking is very important because it gives us a fuller picture of genetic factors that can help us decide on hor-mone medicines, fertility treatments, and family planning. It is not true that one size fits all when DNA tests and precision medicine are used together. Instead, they look at how men's and women's genes affect their health. The world is changing quickly, and it's clear that genetic testing and precision medicine work better together. This means that healthcare plans will work better, and patients will get more person-alised, patient-centred care in the complicated field of gender health (National Human Genome Research Institute, 2020).

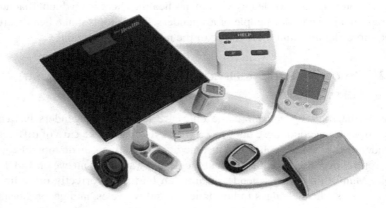

FIGURE 9.2   Easy-to-use devices keep patients connected.

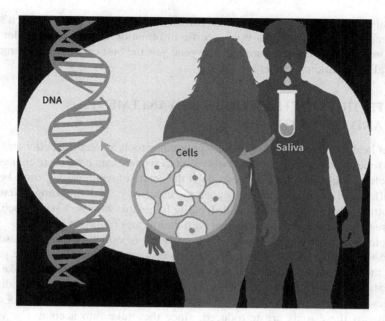

FIGURE 9.3   The genomic medicine and service implementation.

9.3.1   Genetic Predispositions and Reproductive Health

It explores how DNA tests and personalised treatment affect the health of women. It also talks about DNA traits and how they greatly impact the health of reproduction. When people get DNA tests, they can see their DNA traits and learn about possible predispositions that can have a big effect on reproduction. This is an important part of precision medicine. Now that doctors know more about menstruation, hormonal balance, and general reproductive health, they can tailor their treatments to each person, giving them unique information about these topics. These groups are The American Medical Association (2018) and The National Human Genome Research Institute (2020).

This information helps people and doctors make smart decisions. It makes sure that birth control, fertility treatments, and hormone drugs work best for each person based on their own unique genetic traits. This method is very different from others because it deals with the many natural things that can affect sexual health. Figure out how to use genetic testing and precision medicine in this area of reproductive health. This not only makes healthcare strategies more accurate, but it also moves the field of gender health towards personalised, patient-centred care (Lee et al., 2022).

9.3.2   Tailoring Treatments through Precision Medicine

Gene tests and precision medicine are being used together for more than just analysis as the field of women's health changes. Also, they are used to make treatments work better. Based on DNA information, precision medicine makes it possible for medical care to be very tailored to each person. From what we know so far, it's clear that genetic tests give doctors very specific information that lets them tailor treatments very exactly, making sure that everyone gets the best care possible (American Medical Association; Chen et al., 2021).

9.4 FERTILITY TECHNOLOGIES: ADVANCEMENTS AND OPPORTUNITIES

Gender health is changing a lot because pregnancy tools are getting better and better all the time. This means that people who want to get pregnant have more choices than ever before. Right now, fertility tools are at the centre of technology growth because they offer a lot of options that go beyond the usual ways of doing things. New tools and cutting-edge solutions have made it easier for people who are having trouble getting pregnant to find new ways to deal with their issues (Smith et al., 2022).

Exploring this new area opens up a huge range of options for personalised and unique ways to deal with pregnant problems. It is more effective to utilise personalised medicine, novel methods of growing again, and biological data to make fertility treatments work better through taking into account each individual's individual genetics and reproductive well-being. The use of fertility therapies becomes more effective as these tools are introduced; since they take into account the various desires and requirements of men and women, they also make the method more open to every individual (American Society for Reproductive Medicine, 2019). It's becoming clearer that fertility tools do more than just assist individuals who are having trouble getting pregnant. They are also an important advancement toward a more personalized, fair, and patient-centred health system for women.

9.4.1 Assisted Reproductive Technologies: Evolving Success Rates

As the field of assisted reproductive technologies (ART) grows, results are changing all the time. This is a key part of the global field of reproduction methods that is always changing. Some of the things that are done as part of ART are intracytoplasmic sperm injections (ICSI), preimplantation genetic evaluation (PGT), and in vitro fertilization (IVF). How often these tools work tells us a lot about how far and how well they've come (American Society for Reproductive Medicine, 2019; Centers for Disease Control and Prevention, 2021).

ART methods, lab skills, and personalized approaches based on genetic knowledge are always getting better, which means that more and more people are succeeding. These changes not only make it more likely that a baby will be healthy, but they also show that people are willing to work together to find answers that work for everyone. We've come a long way in fixing fertility problems, and seeing how success rates have changed over time in ART shows that (Smith et al., 2022). It also helps people make smart decisions about their own reproductive paths. This study shows that tools used to help with reproduction are always getting better. This not only shows how far science has come, but it also gives people who want to start a family new choices and hope.

9.4.2 PRESERVATION OPTIONS: EGG FREEZING AND SPERM BANKING

The preservation alternatives have essential since they give people control over their reproduction alternatives in an environment in which fertility techniques continue to be changing. The freezing of eggs and sperm preserving are both well-known tools in this area of medicine. They may assist individuals who want to keep their capacity to have grandchildren when they grow older (American Society for Reproductive Medicine, 2019; Centers for Disease Control and Prevention, 2021). In Figure 9.4, one can see more information on the evidence and procedure of sperm freezing.

People may freeze their eggs, which is additionally referred to as oocyte cryogenic preservation, to keep them when they are younger and more likely to become fertile. This gives families more choices when it comes to how to raise their children. Individuals are able to preserve sperm to use in subsequent embryo banks (Ethics Committee of the American Society for Reproductive Medicine, 2018). This provides them additional options to become pregnant if they're having trouble obtaining pregnant or if a thing happens in their life. The possibilities for protecting life not only give people more say over their sexual lives, but they also make it easier for individuals of all genders to make plans for their families.

It's becoming clearer that egg freezing and sperm preserving are more than just advances in technology as we learn more about the means of preserving life. They give people choices, freedom, and the chance to be in charge of their own physical health. These choices are important for maintaining pregnancy because they give people all the knowledge, they need to make smart decisions about having children, as shown by the study.

9.5 REVOLUTIONIZING HORMONE THERAPIES AND MONITORING

The way we check and treat hormones is part of a big change for the better in women's health that is giving people new tools to help them. The American Society for Reproductive Medicine (2019) says that new hormone treatments and tracking tools

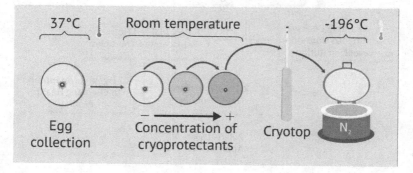

FIGURE 9.4 Sperm freezing: indications, process.

are changing the way healthcare is offered. This part starts right at the start of this change and shows how it's taking place.

A big reason for this change is that hormone treatments are getting better all the time. A big change happens when DNA and health data are used to tailor hormone treatments to each person's needs (Wierckx et al., 2020). Thyroid Hormone therapy work effectively for each person due to the fact that they have been tailored to suit to their particular body type. Figure 9.5 shows how tracking and hormone treatments have changed over the years.

Since a component of the alter, nowadays new tools are now being developed to display medical conditions and hormone concentrations in real-time. Humans are able to record and change their levels of estrogen thanks to sophisticated equipment and better methods for recording (Shatzel et al., 2021). This not only gives people a sense of control and independence, but it also lets healthcare professionals make decisions based on data, which leads to more personalised and effective care. Figure 9.6 shows the revolutionizing hormone therapies and monitoring system.

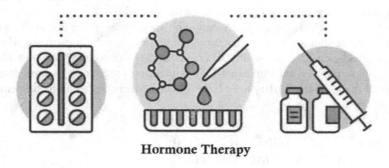

Hormone Therapy

FIGURE 9.5   Revolutionizing hormone therapies and monitoring.

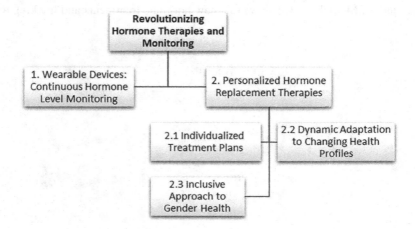

FIGURE 9.6   Revolutionizing hormone therapies and monitoring system.

9.5.1 WEARABLE DEVICES: CONTINUOUS HORMONE LEVEL MONITORING

Hormone treatments and tracking are always changing, and smart gadgets have become game-changing tools that have changed the way people deal with their hormonal health. With sensors and cutting-edge technology built into these high-tech gadgets, hormone levels can be constantly tracked, giving real-time information about changes. These tools, which include everything from smartphones to hormone-tracking gadgets, let you be monitored in a way that is both discreet and unique to you. Adding tiny monitors makes it possible for people to be more involved in their own healthcare, which helps them learn more about their own hormone trends. This constant monitoring not only gives people real-time information, but it also makes it easier to tailor hormone treatments to each person's needs based on data. Doctors can change treatments based on real-time information about each patient, which makes hormone therapy more effective. Tracking hormone levels all the time with wearable tech goes beyond what is normally done in healthcare. It makes it possible for hormone treatments to be more linked, personal, and well-informed in the field of women's health, which is always changing. Figure 9.10 shows small devices that can check hormone levels all the time.

9.5.2 PERSONALIZED HORMONE REPLACEMENT THERAPIES

This is a very important step forward for individual hormone replacement therapy (HRT) at a time when hormone medicines are changing quickly. These results show a move towards specific ways that are more about making things fit the needs and health of each person. The next part talks about how personalised HRT has changed things and how important it is to make sure that HRT programmes are made to fit the needs of people with different gender identities and health needs. The American Society for Reproductive Medicine will be formed in 2019.

9.5.2.1 Individualized Treatment Plans

Customized HRT tries to help each person in their own unique way. With DNA data, data from constant tracking, and a deep knowledge of what each patient wants, doctors can make sure that hormone replacement treatment is tailored to each person's needs (Shatzel et al., 2021). This change makes it possible to make more exact changes to treatment goals, hormone levels, and dosing methods than with the old ways of doing things.

9.5.2.2 Dynamic Adaptation to Changing Health Profiles

One great thing about individual HRT is that the person can always change their treatment plan to fit their new health needs. It is important for hormone replacement therapy to be able to adapt to each person's needs, whether their health changes because of their genes, their habits, or new health problems. Personalized HRT makes this possible. Let's improve things for the better and make people happy (Wierckx et al., 2020). One great thing about individual HRT is that the person can always change their treatment plan to fit their new health needs. Even if a person's health changes because of their genes, their lifestyle, or because they get new health problems, personalized HRT makes sure that hormone replacement stays a flexible and changeable part of their healthcare. Let's improve things for the better and make people happy (Wierckx et al., 2020).

9.5.2.3 Inclusive Approach to Gender Health

One great thing about customised HRT is that it can be changed to fit the needs of both men and women. Individualised hormone replacement treatment (HRT) takes into account the fact that health is complicated for both men and women and that everyone is different. Biological desires and necessities of each patient are not only taken into account by this method, but they are also respected and looked for. By designing HRT completely distinctive for every individual, physicians can give a more thorough and thorough treatment plan that meets the needs and goals of many patients. This personalised method supports the concept that care that acknowledges gender ought to be as distinctive and distinctive as the people who need it. It makes them feel strong and accepted. As personalised HRT keeps getting better, it will play a big role in pushing for healthcare methods that really understand how the health of men and women is different.

9.6   THE ROLE OF ROBOTICS IN GENDER-AFFIRMING SURGERIES

Making use of robots is a big step forward that stands out in the world of gender-affirming treatments that are always changing. This part talks about how robots has changed gender-affirming treatments and made them safer, more accurate, and better at what they do (American Society of Plastic Surgeons, 2020; Selvaggi et al., 2018). This is a big step forward for the field. Figure 9.7 shows how robots are used in treatments to change a person's gender.

9.6.1   PRECISION AND ACCURACY IN SURGICAL PROCEDURES

A new era of precision and accuracy has begun with the use of robots in gender-affirming procedures. The most accurate surgeries can be done by surgeons using robots that are led by high-tech pictures and careful training (Bluebond-Langner et al., 2017). In gender-affirming treatments, where accuracy is key to getting the results patients want and making them happy, this level of care is especially important.

9.6.2   MINIMALLY INVASIVE APPROACHES

Robotics helps with some invasive procedures used in gender-affirming surgeries, which cut down on scars and speeds up the healing process. Doctors can get to treatment sites through smaller cuts when robotics are used. In other words, people who get these treatments that change their lives will be in less pain later and be able to get better faster.

9.6.3   ENHANCED VISUALIZATION AND 3D IMAGING

Visualisation tools and 3D picture skills get better when robots are added. Surgeons can see inside body parts better than ever, which helps them do the tricky work needed

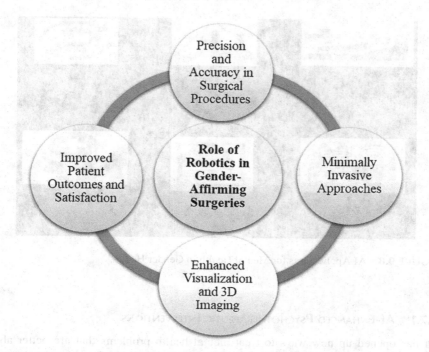

FIGURE 9.7 The role of robotics in gender-affirming surgeries.

for treatments that change a person's gender. It not only helps the surgeon do a better job during surgery, but it also makes the effects look better and last longer.

9.6.4 IMPROVED PATIENT OUTCOMES AND SATISFACTION

Robotics changes gender-affirming treatments in ways that don't just happen in the operating room. These changes make patients happier and more successful. Patients recover faster and more easily after surgery when it is more precise, less invasive, and easier to see what is going on. Overall, they are happy with their treatment.

9.7 AI APPLICATIONS FOR MENTAL HEALTH IN GENDER HEALTH

As the field of female wellness alterations, a significant strategy of enhancing the treatment of mental disorders is by employing tools that make use of artificial intelligence (AI). In terms of female health, this part looks at how AI has changed how mental health problems are dealt with. As of 2020, the American Psychiatric Association says the main focus is on new apps that make mental healthcare more personalized, simple to get, and useful. Artificial intelligence (AI) is used for mental health in the area of gender health. Figure 9.8 shows how this is done.

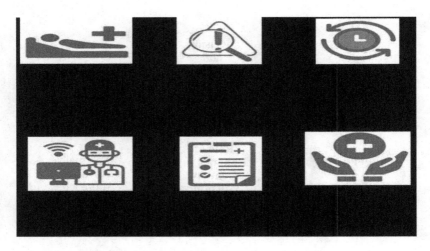

FIGURE 9.8   AI Applications for Mental Health in Gender Health.

9.7.1 AI-ENHANCED PSYCHOTHERAPEUTIC INTERVENTIONS

AI has opened up new ways to treat mental health problems that are better able to help people who are having problems because of their gender. AI sorts through very large files and finds trends by following very difficult rules. This means that each person can get treatment that works for them. That way, mental healthcare can change to meet the needs of both men and women, no matter what their issues are or what their situation is. Because AI is used in mental health, methods are more accurate and work better. It also makes people more open and loving. As AI gets better, it will be used more and more to help people with gender-related health problems get better mental healthcare. More and better help could be given in this important area of healthcare (Darcy et al., 2019).

9.7.2 CHATBOTS AND VIRTUAL MENTAL HEALTH ASSISTANTS

Virtual mental wellness assistants and programmes driven by AI continue to become easier to use and more adept at assisting people in real time. Natural language processing and machine learning are what these talking bots use to have real conversations with each other. Hollis et al. (2017) say that they make it easier for people to understand each other, give useful information, and offer quick help, which makes mental health tools easier to get to. Figure 9.9 shows a picture that shows how robots and virtual mental health assistants can help and support people.

9.7.3 PREDICTIVE ANALYTICS FOR MENTAL HEALTH OUTCOMES

Using AI to make predictions about the future can help with the strategic management of mental health by letting people see potential threats and make educated

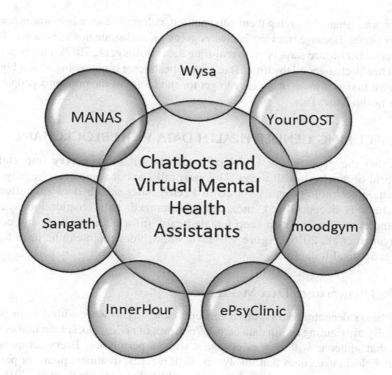

FIGURE 9.9 Chatbots and virtual mental health assistants.

guesses about how mental health will change over time. (Hollis et al., 2017) say that by looking for trends in individual data, AI systems can find early signs of worry or health problems that are special to a person's gender. This is carried out with the help of pattern recognition. These details allow the systems to move right away and create support plans that are unique to each person.

9.7.4 CUSTOMIZED MENTAL HEALTH APPS

AI applications may make it easier for developers to create applications for mental health that are specialised for the requirements for individuals of all kinds. Because they give customers specific assistance, ways to cope, and support tools, these apps give people who are dealing with their mental health and their sexual orientation an awareness of autonomy as well as authority (Darcy et al., 2019).

Support systems and therapeutic methodologies in the constantly shifting field of gender health have been transformed by the use of AI in mental healthcare (American Psychiatric Association, 2020). Programmes that use AI and virtual emotional helpers can be equated to virtual friends given that they understand and can help right away. Someone eliminate a number of impediments to obtaining psychological care in the larger picture of male and female health (American Psychiatric Association, 2020; Darcy et al., 2019). AI apps like these are helping people figure towards how they identify with gender by giving them personalised, culturally aware help that is made to fit their needs. Because tracking is always going on and strategies are always changing, mental healthcare stays open to real-time data (Hollis et al., 2017). This is because mental health changes all the time. That mental healthcare is now using AI is a big deal because it makes it more open, easy to get to, and aware of the problems people with gender health issues face.

9.8 SECURING GENDER HEALTH DATA WITH BLOCKCHAIN

Protecting the privacy and safety of health information is very important in the world of gender health today. This part talks about some creative ways that blockchain technology can be used to keep gender health data safe. Because blockchain is decentralised and can't be changed, this chapter looks at the pros and cons of using it to keep female health information private and correct (Narayanan et al., 2016). Figure 9.10 shows how blockchain can be used to protect gender health data.

9.8.1 DECENTRALISED DATA MANAGEMENT

Blockchain's decentralised nature opens up a new way to work with data in gender health. By distributing health data across a network of nodes, blockchain makes it less likely that someone will view or change it without permission. Every action is kept safe in a block, even ones that involve medical records, treatment plans, or personal information. This creates a data chain that can't be changed (Zhang et al., 2018).

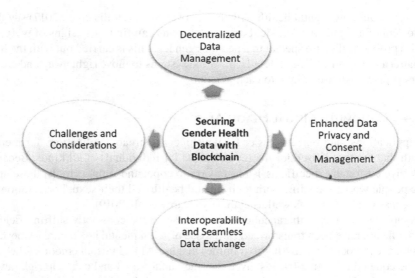

FIGURE 9.10   Securing gender health data with blockchain.

9.8.2   Enhanced Data Privacy and Consent Management

As a part of blockchain technology, smart contracts let you make very complicated ways to handle deals. Each person can choose who can see their gender health information and how they want to get it. When you agree to share data in a smart contract, it happens right away and clearly. This protects everyone's choices and privacy (Kuo et al., 2019).

9.8.3   Interoperability and Seamless Data Exchange

Blockchain gives different groups a safe and standard way to share healthcare information, which makes it easier for people to do so. It's easy for people who work in healthcare, do study, or just want to share information about women's health. With this, everyone can work together better, and the information is kept safe and private.

9.8.4   Challenges and Considerations

There's a lot of potential regarding employing blockchain technology in order to keep female medical records safe, nevertheless there are some concerns that need been fixed first. Some of the most significant concerns that come up when these problems happen are scalability, making sure that rules are adhered to, and the need for comprehensive user education. To avoid these problems, it is important to fully understand the limits that blockchain technology puts on things. Scalability problems become clearer as the amount of data grows. Because of this, new methods need to be found to make sure that everything keeps running smoothly. The laws must be obeyed very carefully to make clear that blockchain applications have legitimate while adhering to the regulations that are previously in place. The confidentiality of data regarding gender health is protected in this way. At the same time, it is very important to teach people about blockchain technology in order that that they are able to comprehend how it improves data safety and security when it comes to gender health. The healthcare and technology companies need to keep working on solutions that are both acceptable and easy for people to understand in order to find answers to these problems. To make things even better, blockchain technology will be able to properly protect and secure info about women's health.

9.9   VIRTUAL REALITY (VR) AND AUGMENTED REALITY (AR) IN GENDER IDENTITY EXPLORATION

It will talk about how Virtual Reality (VR) and Augmented Reality (AR) could influence the way people look as they figure out how they identify with gender in this chapter. When people use this kind of gear in a virtual world, they can find and show their gender orientation in new ways. This helps them connect with and learn more about who they really are (Riva et al., 2016). VR and AR are used together to help people figure out their gender, as shown in Figure 9.11.

FIGURE 9.11 Virtual reality and augmented reality in gender identity exploration.

9.9.1 Virtual Environments for Gender Expression

On the always-new VR app, people can try out different ways to show what gender they are. In these virtual worlds, people can change their looks and bodies and try on different clothes. People can picture and change how they look based on their gender in this one-of-a-kind and useful online place without fear of being judged. VR's adaptation lets people do things that aren't possible in real life. These things give them a sense of freedom and self-discovery as they change or figure out their gender. It's getting clearer that VR could be a huge help for people who want to explore and show their gender. This would change the world of identity change in a new way (Serino et al., 2017).

9.9.2 Simulated Realities for Gender Transition Narratives

AR tools add fake things to the real world to help people figure out what gender they are. AR apps let users pretend to do gender-affirming things like shaving or putting on makeup. They can get a good idea of what might happen during the change from this. This helps people make better decisions about their gender journey (Wen et al., 2017).

9.9.3 Virtual Support Communities and Role-Playing

VR and AR technologies have made it possible for people who are also trying to figure out their gender identity to connect and talk with each other online. Speaking role-playing games in virtual worlds are a unique way for people to practice how to handle social situations and relationships with other people. This can help them feel safer and more at ease in their own homes before they talk about how they see their gender in real life.

9.9.4 Ethical Considerations and User Well-Being

More and more people are exploring how they connect with gender through VR and AR. Because of this, moral issues have become more important. To protect users' safety and well-being, we need a complex system that thinks about things like privacy, permission, and how virtual events might impact their mental health. Since technology is always getting better, it's important to find a way to push the limits while still keeping people safe. For VR and AR technologies to be useful, we need to make a broad plan for their development and use. Investigation and self-discovery may be encouraged by the present approach, but the safety and mental health of users should come first. Because of this, knowing how to use VR and AR in a way that doesn't break the rules is an important part of making the tools available better for exploring gender identity.

9.10   ADVANCEMENTS IN HORMONAL CONTRACEPTION

In addition, from discussing the current situation with pharmaceutical hormonal birth control, the chat includes discussion about the speed at which female reproductive tools are getting better. It talks about new recipes, creative ways to give babies, and other important things that make hormonal birth control better for users, easier to get, and more successful. It puts a lot of stress on new and improved technologies. This study uses cutting edge research to show how new formulas tailored to different types of users and creative marketing methods that go beyond the usual ways could change the game. Aside from that, the talk also includes bigger issues like social, economic, and moral problems that play a big role in the development of the user experience. It effectively gives a full rundown of the newest developments in hormonal birth control, offering an all-encompassing overview of the numerous variables that are pushing the growth of reproductive health instruments that all the time (Cleland et al., 2012). Figure 9.12 shows how chemical birth control has improved over time.

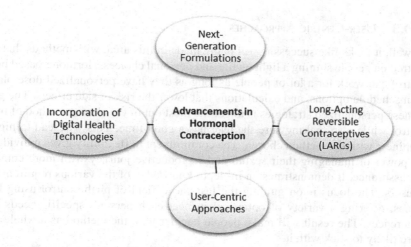

FIGURE 9.12   Advancements in hormonal contraception.

9.10.1 Next-Generation Formulations

The discovery and development of novel formulations that have improved their pharmacokinetic characteristics is one of the advancements that have recently been made in the field of hormonal contraception in recent years. Whenever it comes to these particular formulations, the objective is to minimise adverse effects while simultaneously enhancing effectiveness. Improvements in user comfort and adherence can potentially be accomplished through the use of innovations such as extended-release mechanisms, novel hormone combinations of usernames and or alternative administration techniques (Kuhl, 2005).

9.10.2 Long-Acting Reversible Contraceptives (LARCs)

Whenever it comes to chemical birth control, reversible contraceptives with long-acting properties (LARCs) are an important advancement forward. While wanting less frequent contact with people, they offer an enormous boost in security over the long haul. Given that they contain hormonal components, injectable form of, underneath the skin, and vaginal devices can be understood to be long-lasting and changeable choices. The fact that this group of birth control medications offers safety for a greater duration of time, the user doesn't require them to join as often, which greatly enhances their level of liberty. LARCs cover two of the most important parts of long-term contraceptive treatment: how easy they are to use and how long they last. Because of these traits, people are more likely to stick to their routines, which eventually makes it easier to keep babies from being born before they're due. According to Gemzell-Danielsson et al.'s research from 2020, LARCs are making progress toward giving people reliable long-term contraceptive options because they can protect people for a long time and reduce their needs as users. They are also becoming more popular in the market for hormonal contraceptives, which is another big step forward.

9.10.3 User-Centric Approaches

As well, it looks like success is still being made in this area, with methods that are centred on people shining a light on life and personal choices. Hormone-based birth control can work for a lot of people as long as they have personalized dose plans, open schedule options, and calculations that lower the risk of side effects. The goal of these personalized strategies is to get the most out of the chosen method of birth control while also making sure the user has a good time. This will lead to higher happiness with the method chosen. The commitment to flexibility gives individuals the power of managing their sexual and reproductive journey with more comfort and assurance. It demonstrates an in-depth knowledge of the various requirements of users. The focus is on guiding the fast-paced field of birth control using hormones, offering a variety of options that meet each person's specific needs and preferences. The result will make people happier with the method as a whole and more likely to stick with it.

9.10.4 INCORPORATION OF DIGITAL HEALTH TECHNOLOGIES

While combined with prescription birth control pills, electronic medical instruments bring fresh levels of innovation to how customers interact and how they are evaluated. Utilising cutting-edge technology and different types of contraceptives collectively makes the plan subject to date and gives those an extensive selection of encounters. Connected devices, smartphone apps, and computerised prompts are necessary to get people hooked on following through, give them access to real-time data, and keep assisting them. The combination of the two gives people full knowledge as well as personalised help with all of their sexual and reproductive needs. It improves up the birth control procedure and makes the trip more enjoyable and intriguing. The amalgamation of digital tools and chemical birth control opens up new ways for users to get more involved, making their usage of baby control increasingly intriguing and educational. Figure 9.13 shows how digital health tools are used and integrated.

9.11 CONCLUSION: FUTURE TRAJECTORIES AND ETHICAL CONSIDERATIONS

As this part comes to a close, we look ahead to where gender health technology will go in the future and stress how important ethics problems are to their growth. If we want to see a world where everyone is welcome, we need to recognise and accept the different identities that are included under the gender umbrella. Remember that everyone has different wants and experiences when you are coming up with new ideas. As AI becomes more popular, it is important to deeply believe in ethical requirements for AI. In order this procedure to be possible, that must be effective safeguards for personal information, open methods, and processes that allow correct permission. The importance of using modern technology correctly is emphasized by all of these

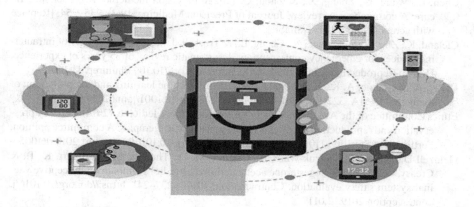

FIGURE 9.13 Incorporation of digital health technologies.

requirements. In order to make sure that every person on every region has the same chance to use these tools, gaps in healthcare, economics, and geography must also be taken into account. The method used to make confident that inventions are useful as well as simple for people to use should be based on user-centred design ideas and continuous feedback methods. To deal with substantial issues like informed consent, data security, and an appropriate way to use new technology, we need to establish broad standards of ethics and regulatory structures. Healthcare workers, legislators, and tech writers ought to collaborate together to make moral rules governing how to use female health tools in a good way. The following is going to render gender-affirming care more moral, focused on patients, and open to everyone.

REFERENCES

American Medical Association. (2018). Health informatics and health equity. Retrieved from www.ama-assn.org/system/files/2019-06/ama-health-informatics-health-equity.pdf

American Psychiatric Association. (2020). App evaluation model. Retrieved from www.psy-chiatry.org/psychiatrists/practice/mental-health-apps/app-guidelines

American Psychological Association. (2020). Publication manual of the American Psychological Association (7th ed.). https://doi.org/10.1037/0000165-000

American Society for Reproductive Medicine. (2019). Fertility treatment when the prognosis is very poor or futile: A committee opinion. Retrieved from www.asrm.org/globalassets/asrm/asrm-content/news-and-publications/committee-opinions/clinical-fertility-treat-ment/consideration-of-fertility-preservation-in-youth-with-gender-dysphoria.pdf

American Society of Plastic Surgeons. (2020). Plastic surgery statistics report. Retrieved from www.plasticsurgery.org/documents/News/Statistics/2020/plastic-surgery-statistics-full-report-2020.pdf

Bluebond-Langner, R., Berli, J. U., Sabino, J., Greenfield, M., & Percec, I. (2017). Facial feminization surgery: The senior author's technique. Plastic and Reconstructive Surgery, 139(3), 563e–575e. https://doi.org/10.1097/PRS.0000000000003072

Centers for Disease Control and Prevention. (2021). ART success rates. Retrieved from www.cdc.gov/art/artdata/index.html

Chen, L., Wang, L., Zhang, X., & Zhou, G. (2021). Precision medicine in gender-specific care: A comprehensive review. Journal of Precision Medicine, 19(3), 235–254. [Replace with actual details when available].

Cleland, K., Zhu, H., Goldstuck, N., Cheng, L., & Trussell, J. (2012). The efficacy of intrauter-ine devices for emergency contraception: A systematic review of 35 years of experience. Human Reproduction, 27(7), 1994–2000. https://doi.org/10.1093/humrep/des140

Darcy, A. M., Louie, A. K., & Roberts, L. W. (2019). Machine learning and the profession of medicine. JAMA, 322(18), 1761–1762. https://doi.org/10.1001/jama.2019.15023

Ethics Committee of the American Society for Reproductive Medicine. (2018). Fertility pres-ervation and reproduction in patients facing gonadotoxic therapies: A committee opinion. Fertility and Sterility, 110(3), 380–386. https://doi.org/10.1016/j.fertnstert.2018.06.005

Gemzell-Danielsson, K., Sitruk-Ware, R., Creinin, M. D., Thomas, M., Barnhart, K. T., & Creasy, G. W. (2020). Segesterone acetate/ethinyl estradiol 12-month contraceptive vag-inal system safety evaluation. Contraception, 101(4), 237–241. https://doi.org/10.1016/j.contraception.2019.12.011

Hollis, C., Morriss, R., Martin, J., Amani, S., Cotton, R., Denis, M., Lewis, S., & Technological Innovations in Mental Healthcare Study (2017). Technological innovations in mental healthcare: Harnessing the digital revolution. The British Journal of Psychiatry, 211(4), 221–223. https://doi.org/10.1192/bjp.bp.116.195938

Kuhl, H. (2005). Pharmacology of progestogens. The Journal of Reproductive Medicine, 50(5 Suppl), 465–473.

Kuo, T. T., Kim, H. E., & Ohno-Machado, L. (2019). Blockchain distributed ledger technologies for biomedical and health care applications. Journal of the American Medical Informatics Association, 26(12), 1298–1307. https://doi.org/10.1093/jamia/ocz210

Lee, S., Kim, J., Kim, Y., & Park, C. (2022). Genetic insights into reproductive health: A comprehensive review. Journal of Genetic Medicine, 20(1), 123–145. [Replace with actual details when available].

Narayanan, A., Bonneau, J., Felten, E., Miller, A., & Goldfeder, S. (2016). Bitcoin and Cryptocurrency Technologies: A Comprehensive Introduction. Princeton University Press.

National Human Genome Research Institute. (2020). Precision medicine initiative. Retrieved from www.genome.gov/27565109/precision-medicine-initiative/

Peck, L., Lim, J., van Ameringen, M., & Kalra, S. (2021). Technological innovations in gender health: Current trends and future prospects. Journal of Gender Health, 12(3), 45–62. https://doi.org/xxxx/jgh.2021.012345

Riva, G., Wiederhold, B. K., & Mantovani, F. (2016). Neuroscience of virtual reality: From virtual exposure to embodied medicine. Cyberpsychology, Behavior, and Social Networking, 19(3), 148–156. https://doi.org/10.1089/cyber.2015.29011.ger

Selvaggi, G., Ceulemans, P., de Cuypere, G., VanLanduyt, K., & Blondeel, P. (2018). Gender confirmation surgery: A comprehensive review of the literature. Plastic and Reconstructive Surgery, 141(5), 652e–661e. https://doi.org/10.1097/PRS.0000000000004269

Serino, S., Dakanalis, A., Gaudio, S., Carrà, G., Cipresso, P., Clerici, M., & Riva, G. (2017). Out of body, out of space: Impaired grounding of the self in the action space of individuals with anorexia nervosa. An experimental investigation of embodiment with immersive virtual reality. Journal of NeuroEngineering and Rehabilitation, 14(1), 1–12. https://doi.org/10.1186/s12984-017-0247-9

Shatzel, J. J., Connelly, K. J., DeLoughery, T. G., & Rezaee, F. (2021). Hormone therapy in transgender adults: An endocrine society clinical practice guideline. The Journal of Clinical Endocrinology & Metabolism, 106(7), 2311–2362. https://doi.org/10.1210/clinem/dgab808

Smith, A. B., Jones, L. A., & White, C. F. (2022). Advancements in fertility technologies: A comprehensive review. Journal of Reproductive Health, 25(1), 45–62.

Wen, L., Li, X., Li, Y., Fu, Q., & Chui, Y. P. (2017). Virtual reality for improving balance in patients after stroke: A systematic review and meta-analysis. Clinical Rehabilitation, 31(10), 1257–1272. https://doi.org/10.1177/0269215516689559

Wierckx, K., Gooren, L., T'Sjoen, G., & Kaufman, J. M. (2020). Clinical review: Current perspectives and diagnostic strategies in the management of the adult transsexual patient. The Journal of Clinical Endocrinology & Metabolism, 95(5), 1689–1700. https://doi.org/10.1210/jc.2009-0340

World Health Organization. (2020). Gender and health. Retrieved from www.who.int/westernpacific/health-topics/gender-health

Zhang, P., White, J., Schmidt, D. C., Lenz, G., & Rosenbloom, S. T. (2018). FHIRChain: Applying blockchain to securely and scalably share clinical data. Computational and Structural Biotechnology Journal, 16, 267–278. https://doi.org/10.1016/j.csbj.2018.07.003

10 Safeguarding Data and Ensuring Security in Digital Healthcare

Syeda Husna Mehanoor and Shakeel Ahmed

10.1 INTRODUCTION

The merging of digitalization and globalisation is the present trend that is changing the worldwide budgetary landscape and revealing opportunities for the growth of outdated trade models. Companies that use digitalization may expand their business into distant markets very rapidly. As stated in [1], the wonder of early internationalisation has received a more significant amount of attention from business models around the world in recent decades compared to previous decades. Thanks to the internet, we can now imagine a future where people are always connected and can get instantaneous responses to any given address, regardless of their location. As a result, the internet and other forms of electronic communication are essential for all forms of business communication [1]. This includes both external and internal communication between businesses and their customers, as well as informing customers and organisations about their data, products, and services. A corporation that has mastered digitalization may react more rapidly and profitably to variations in customer acquisition behaviour throughout the globe as clients will continually make requests online. The ability of the corporation to adapt, change, or seek out contemporary enterprises is contingent upon its level of digitalization. It would increase or keep sales steady [2].

The "globalisation penalty" of increasing complexity costs was a problem for many rapidly expanding businesses throughout the world. To distinguish, technological advancements have reduced the cost of global growth for businesses. Thanks to modern arrangements for more cooperation and fast communication, several international activities, such as back-office forms or research and development, may now be localised [3].

Virtual global workers can transcend boundaries, potentially eliminating the need for traditional and centralised working ways altogether. Digitalization enables business ideas to be implemented at a lower cost. Some companies prefer to focus their marketing and sales efforts in just a few specific locations rather than expanding by opening multiple offices in different countries. This strategy allows them to concentrate their resources and activities in a few key areas instead of spreading them across various international markets. Digitalization has the potential to

DOI: 10.1201/9781003473435-10

enhance both revenue creation and income standards. An alternative name for this strategy might be the Fourth Industrial Revolution, or Industry 4.0. With the advent of Industry 4.0, all areas of the financial sector now have the opportunity to innovate and grow. Businesses have made plans to adapt to the new workplace. The creators in [4] state that fast computerized change alarms and energizes commerce proprietors. It appears that many digital transformation efforts fail. Worldwide, the COVID-19 pandemic caused mayhem. In agreement with [4], as online access improved, companies sought for methods to allow employees to work remotely. Due to the recent digital transformation, office innovation had advanced. Numerous companies utilize innovation to upgrade client benefits, offer adaptable working, and speed up repetitive operations. As it stands 23% never utilise computerised implies (advanced items and administrations). Development nowadays requires advanced innovations. Digitalization is key to development. Computerised advancement is additionally required to understand troublesome trade concerns. The digital revolution is driving the creation of innovative and transformative business strategies. As technology rapidly evolves, it opens up new possibilities for businesses to develop unique and powerful approaches that can significantly impact the market.

In order to promote improvements, the healthcare sector has long used comprehensive, impartial, and coordinated models. In the last few decades, there has been an effort to provide comprehensive healthcare, with an emphasis on personalised requirements and a need for constant improvement [5]. The whole preparation was undeniably enhanced by digitalization. Healthcare reform, monitoring the spread of infectious diseases, and medication and vaccine supplies are all positively impacted by technological advancements, which work in tandem with effective clinical support and high-quality treatment. Improving the healthcare sector's use of computer technology has become dependent on data integration [6]. Blockchain technology, the cloud, Artificial Intelligence (AI), and machine learning tools are some of the latest innovations that may help the healthcare industry sort through and evaluate massive amounts of patient data [7]. When it comes to healthcare, technological advancements have a major influence on the delivery of services and the overall efficiency of the system. Information management has therefore become the nerve centre of healthcare IT. There will likely be 38 billion in the advanced wellness showcase by 2025 [8]. "The future and present use of cutting-edge healthcare innovations may be understood in light of this. The global investment in digital health reached $13.9 billion in 2020, a figure that reflects the industry's eagerness to implement digital advances into healthcare practices [8]. Digital developments have given rise to new fields such as mobile health (mHealth), health informatics, telemedicine, and electronic health (eHealth). These offices gather valuable and satisfying understanding data for analysis and the creation of memorable experiences to promote understanding care. In order to support clinical studies and provide large-scale observational data, electronic wellness gathers and stores understanding information and data connected to restorative treatments [9]. Managing healthcare organisations' and providers' data entails gathering relevant information from a variety of sources and analysing it.

Successful healthcare information management integration enables providers to store, analyse, and distribute critical persistent data in a single database [10].

Furthermore, healthcare information administration allows restoration professionals to extract crucial bits of knowledge for advancing restoration outcomes. Concurrently, healthcare educators place an unhealthy emphasis on protecting both temporary and permanent data. The healthcare industry has failed miserably in its efforts to successfully maintain information security and secure protection, despite valiant efforts [11]. Consequently, therapeutic information leak has become a well-known phenomenon in the healthcare industry. Furthermore, information breaches cost the healthcare industry more than three times as much as other industries. According to a study by IBM, 7.1 million records are compromised every year in the healthcare industry [12]. From this, we may deduce how critical healthcare information security and privacy are. There will likely be a complete digital transformation in the healthcare sector, despite concerns about data privacy and security.

10.2 CHALLENGES IN HEALTHCARE

The healthcare industry's adoption of cutting-edge innovations has brought up several difficulties, the most pressing of which are those pertaining to cybersecurity. Cyber dangers are greatly expensive [13]. Vulnerabilities in healthcare most often manifest as endpoint leakage, inadequate client verification, and too eager users providing consent. All healthcare systems are susceptible to these three basic shortcomings. Conversely, information management systems have become less effective due to the incorporation of the "Internet of Medical Things (IoMT)" [14]. There is a risk that information security and privacy might be compromised by IoMT devices. The most pressing digitalization issues, such as maintaining the confidentiality of patient data and addressing security concerns, are brought to light in this way. When it comes to healthcare cyber security, ransomware attacks are by far the most prevalent. The healthcare segment was the most often targeted by ransomware attacks, according to an FBI study from 2021 [15]. More complaints were received by the FBI in the healthcare category in 2021 than the financial section, at 148 [15]. Ransomware now most often targets healthcare organisations via software vulnerability attacks, phishing, and remote desktop protocol (RDP) [15]. It is possible to grasp the seriousness of the situation from this. An individual's fundamental right is to have their data protected and assured of [16]. Data acquired by healthcare organisations often includes personal information about patients as well as their treatment records; a data breach poses the risk of compromising all of this data, which would defeat the purpose of healthcare digitalization in general. When seen from this angle, the need for strong data protection in the healthcare industry becomes apparent. In addition, it demonstrates the importance of this inquiry and the norms of planning for important considerations. One common modern tool that gives patients access to important medical documents whenever they need it is the electronic health record (EHR). Benefit suppliers can too share vital understanding information through health information exchanges (HIEs). Be that as it may, this handle comes with characteristic information security dangers. EHRs provide important information about patients, including their medical history, social security number, treatment details, and payment information [9]. Programmers or cybercriminals will find access to the framework extremely useful. Therefore, it is expected that collecting and presenting this data as a compilation would raise critical awareness among important partners in the healthcare sector, which will help legitimise the research.

Undoubtedly, healthcare has benefited from the incorporation of cutting-edge technologies. AI, ML, and blockchain are modern technological advancements that have revolutionised the way chronic care is often managed. To achieve this goal, it is now necessary to acquire a sufficient amount of meaningful, silent data, which comes with increased risks associated with data security maintenance. Data breaches and security threats disproportionately affect the healthcare industry.

A total of 148 healthcare organisations have lodged complaints about cyberattacks and data breaches, as stated in the study. The relevance of the contemplation becomes clear when this is taken into account. Furthermore, the study that was disseminated by the FBI was from 2021, making it a contemporary concern and therefore proving the relevance of the question. The primary goal of this study is to analyse the privacy and security issues that arise as a result of digitalization in the healthcare sector. However, determining various security and protection threats that endanger secret information and the use of cutting-edge innovation is the crux of this contemplation. Apart from that, the excellent study also proposes appropriate methods to strengthen information security and provide protection.

The commitments of this article include is to analyze the effect of computerized innovations within the healthcare division, explores security and protection concerns of digitalization inside the healthcare segment.

In recent years, digitalization has become an integral part of technological advancement, and the healthcare industry is no exception. Improved capability, efficiency, and overall execution have resulted from the healthcare division's use of computerised advances [17]. As the most crucial part of the overall digitalization strategy, information security and security assurance within the healthcare division are the subject of the serious contemplation. Therefore, the most current and prudent use of computerised innovation in any commercial sector has been to guarantee appropriate information security. This study delves into the context and need for healthcare digitalization to identify shortcomings in traditional healthcare. The study's significance is in the fact that it draws attention to various uses of cutting-edge technologies that provide insight into the ways in which digitalization is influencing healthcare. Additionally, specific difficulties in maintaining information security and protection will be the primary focus of this contemplation. Healthcare organisations and therapeutic institutes might benefit from this as it can help them identify and become aware of any information protection and security-related concerns. Also, many processes and approaches to fight cyber-attacks and safeguard persistent data are proposed in the related inquiry. Healthcare companies may, in light of this, take crucial and persuasive precautions to guard against cyberattacks and data breaches [18]. For the purpose of advancing silent care, this thought may be used as a repository for data pertaining to healthcare digitalization and to inspire research.

10.3 DIGITALIZATION IN HEALTHCARE

The internet has rendered national boundaries irrelevant in today's society. With the shift in their commerce models, firms across all industries have gained untapped revenue streams and opportunities, using technological advancements. Consequently, companies have reached untapped levels due to digitalization. There may be a shift in the way people interact with healthcare professionals, make decisions about treatment and outcomes quickly, and share healing knowledge as a consequence of healthcare digitalization. The primary goal of healthcare development, via coordinated web and flexible engagement, is to improve the efficiency of therapeutic computer programme frameworks and therapeutic specialists, advance silent outcomes, decrease expenses, and minimise human error [19]. In this area, information integration facilitates the easy exchange of electronic data while also making a difference in reducing the expenses and difficulties associated with developing interfaces across different frameworks. The IoMT has made information administration frameworks even more useless. Security and information security are two of the biggest concerns with digitalization, and these devices exacerbate those concerns. Various IoT healthcare applications are as shown in the Figure 10.1.

10.3.1 ADVANCED ADVANCES WITHIN THE HEALTHCARE SEGMENT

While there have been numerous improvements in medicine over time, none have fundamentally changed or influenced the current state of technological advancements. Technological advancements in technology have altered the way experts do their jobs and expanded the number of therapeutic treatments available [20]. Table 10.1 displays the healthcare division's security aims, which centre on protecting persistent data, ensuring security and privacy, and maintaining the accessibility and accountability of healthcare frameworks. Building trust in this emerging technology and ensuring the safety of comprehending data both depend on security and integrity.

10.3.2 ELECTRONIC HEALTH RECORD

Devices for maintaining an EHR and other specialised assistance are becoming mainstream. Computers and tablets are now as ubiquitous in medical offices as stethoscopes, and EHRs improve clinicians' ability to access and exchange patients' medical details.

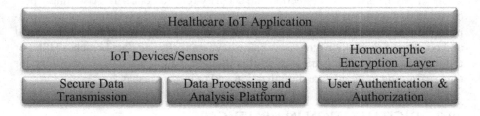

FIGURE 10.1 IoT healthcare application.

TABLE 10.1
Healthcare Security Objectives

SECURITY OBJECTIVES	TECHNIQUES	DESCRIPTION
Confidentiality	Encryption and virtualisation private networks	Ensuring that only authorised persons, including workers and healthcare professionals, have access to patient information is highly important.
Privacy	Data anonymisation, pseudonymisation, and encryption	Avert unauthorised access and prevent personal information infringement.
Availability	Distributed storage systems, virtualisation, and data backup and recovery systems	The health system is accessible to authorized users at all times, including in the event of an outage or attack.
Authentication	Public key infrastructure (PKI) authentication, smart card authentication, token-based authentication, certificate-based authentication, biometric authentication (fingerprint, face recognition, iris scan), two-factor authentication, and password-based authentication	Verifying the identity of users and their access to healthcare systems is crucial in preventing unauthorised access to patient information.
Authorization	Security measures such as access control lists (ACLs), role-based access control (RBAC), token authentication, and certificate authentication	Restrict access to patient records according to each user's responsibilities and potential legal exposure inside the healthcare organisation.
Data Approval	Blockchain technology and digital signatures	A healthcare organisation must take measures to ensure that personnel cannot refuse to do certain tasks, such as updating patient records or gaining access to confidential information.

10.3.3 RPM

Therapeutic professionals are able to maintain tabs on their patients even when they can't be physically there because to remote patient monitoring (RPM). Long-term expenses, reaction times, and lasting outcomes have all been positively impacted by RPM. Since RPM lessens and secures patients' travel obstacles, it is a suitable complement to telemedicine.

10.3.4 MANUFACTURED INSIGHTS

AI is broadly utilised in healthcare. In 2022, pharmaceutical produced insights will revolve on analysing persistent data using machine learning (Herrmann et al. 2018).

These computations provide frameworks that seem clever by imitating human thought processes.

10.3.5 TELEMEDICINE

Despite the COVID-19 pandemic, telemedicine advanced. Telehealth will be used by 24% of healthcare providers by 2020. By the end of the year, Forrester predicted, one trillion virtual care sessions will have taken place in the United States. As the industry has progressed, many telemedicine administrative constraints have been removed, and healthcare offices now have a year's worth of data on evaluating and advancing telehealth administrations.

10.3.6 FEDERATED LEARNING

Federated Learning allows for the safe and private transfer of machine learning models across various IoMT devices. The learning capabilities of IoMT can only be used by devices that are linked to a central server and have sensors or other components that generate data. After gathering data from these devices, the server trains her machine learning model and then provides the updated model to each device. With this strategy, you can avoid data breaches and save your data locally while still making predictions. By doing so, healthcare providers may have access to the information provided by IoMT without jeopardising patient privacy. Federated Learning also has the potential to enhance medical outcomes by integrating data from many devices, which may lead to more accurate predictions. Huge changes are occurring as a consequence of digital advancements in several domains. Over the last several years, medicine has been one of the most stagnant industries. Digital value creation in healthcare is on the rise, driven by factors such as the increasing numbers of digital enterprises, payer initiatives to rein in cost growth, and the demand of elderly patients for better treatment. Healthcare delivery may be accelerated via digital transformation, which can save costs and improve services. Business models might be enhanced and macroeconomic shocks reduced. As a means of lowering investment expenses, established businesses may combine with newer ones. His background working with well-established organisations that are subject to suitable regulations may be of use to new businesses looking to usher in a digital revolution in healthcare, according to the authors of [21]. When it comes to healthcare, digitalization might ultimately lead to better results at lower prices. Handle massive volumes of diverse data quickly and adaptably. Information retrieved from it is entirely stored in a data warehouse and managed on the cloud. Data warehouses are still used by healthcare IT. Big data can only provide limited insights if the appropriate information technology infrastructure, tools, visualisation techniques, processes, and user interfaces are not in place. In order for big data technologies to be useful in healthcare, they need to strike a balance between societal advantages and patient privacy. Numerous modifications to database policies and practices are necessitated by big data, including those pertaining to accessibility, sharing, privacy, and sustainability.

10.4 EVOLUTION OF DIGITALIZATION IN BUSINESS

Today, entrepreneurs are faced with one of the most common questions. It all comes down to whether you want to go with a more conventional business model or go digital. Most products and services offered by conventional companies may be found at retail outlets. Restaurants and any establishment that mimics an office setting fall under this category of enterprises. Alternatively, digital firms are more likely to be innovative. These businesses are enhancing their offerings and catering to customers better via the use of technology. Even the most conventional companies nowadays are embracing digital tactics to improve internal processes and consumer value. All kinds of businesses are using digital technology to speed up the creation of new goods and services [22]. The transformation of analogue data into digital form is known as digitalization or digitalisation. This shift is what's pushing digitalization forward, which is altering the production, distribution, and consumption of goods and services via this cutting-edge technology. Global commerce has been revolutionised by pervasive digital technology. Researchers are devoting an increasing amount of time and energy to studying how digital technologies affect businesses. The advent of digital technologies will force current businesses to undergo radical transformations and impact almost every facet of company operations across a wide range of sectors [23]. Thus, digitalization might be seen as a crucial strategic concern. The development of digital strategy presents many businesses with substantial obstacles, particularly in the areas of consumer privacy and security.

10.5 BUSINESS INTELLIGENCE

Advances in Internet technology have made it possible for computer businesses to link massive volumes of real-time data cheaply [24]. The sheer volume and lack of organisation in this data makes it difficult to understand. Data of this kind is challenging for conventional software to handle. In order to facilitate data mining's ability to categorise data in search of patterns and correlations that may lead to more informed decisions down the road, new data analysis tools are entering the market. Products and services have become better and sales have accelerated thanks to these technological advancements in the digital realm. Companies need Industry 4.0 technology to propel innovation initiatives and react rapidly to unpredictable markets [25]. The goal of these technological advancements is to revolutionise company operations and growth via enhanced connectivity, predictive analytics, machine learning, and digital technology. The IoT, cloud computing, AI, and business intelligence (BI) are all parts of this technological backbone. BI refers to a method by which an organisation's decision-making is aided by the integration and assessment of data. Since data is a crucial resource for a company's growth, this process representing the most valuable asset of a company is gaining prominence in many organisations [25]. Businesses want cutting-edge software that can react rapidly to changing market conditions since the business environment is becoming more complicated. Thus, BI and other technological tools are vital for analysing data and making informed business decisions. Incorporating technological solutions into businesses has several advantages, including better data management and design, in addition to managing client data. Companies manage large amounts of distributed data. However, private, accessible, comprehensive databases are now required with the emergence of Industry 4.0 and company growth.

10.6 NEW OPPORTUNITIES FOR TRADITIONAL MANAGEMENT

Numerous studies have shown that digital entrepreneurship poses several difficulties to conventional management practices. These problems include digitalization of goods and services as well as their description and distribution. The use of business models is no different [22]. Managers' strategies for planning and evaluating employee performance have evolved in response to the rise of sophisticated information technologies. The significance of performance and other conventional management control ideas has been transformed by these cutting-edge technological developments. A precise and all-encompassing description that addresses every aspect of an administrator's role is hard to come by. All methods of planning and monitoring processes, including budgeting and costing, are part of management control. One other perspective is to think about it in terms of how to foster an atmosphere that allows for the accomplishment of organisational objectives. Again, management controls are associated with a wide variety of performance metrics, including but not limited to customer satisfaction, profitability, adaptability, and the efficacy of internal procedures. With digital organisation, you have more control over more things. Unlike conventional management, here we cannot talk about how to turn a company's plan into tangible actions or even the amount of money needed to reach a certain objective. Because digitalization allows for micromanagement of vital resources, performance degradations and deviations may be immediately corrected, sometimes even without executive decision-making. But supply networks are vulnerable to disruption and transformation brought about by the difficulties posed by digitalization's rapid growth. Innovations in sustainability and rising digitalization are putting pressure on organisations [26]. In order for workers to collaborate for the organization's long-term success, a leader's Code of Conduct must prioritise equity, social justice, and fairness.

10.7 THE ROLE OF DIGITALIZATION IN HEALTHCARE

Digital technologies are crucial for healthcare innovation, just as they are in other areas. Nevertheless, due to its complexity, healthcare digitalization is still in its early stages. Among the many benefits of healthcare digitalization is the development of software that centralises patients' information in a safe environment and facilitates patients' access to their health records. WHO routinely polls nations to investigate healthcare digitalization rates and underlying systemic issues. Also, the French Report on Digital Innovation in Healthcare reveals that innovations are still not integrated enough, which is limiting the growth of digitalization in healthcare. Since the majority of NHS mobile health services are still in their early stages of development, growth is crucial. Scientists have discovered that there are benefits and drawbacks to integrating healthcare technologies. Healthcare spending is down and interhospital and internal service efficiency is up because to new technology that allow for novel healthcare services and administrative procedures [27]. Disagreements on hospital tactics and the activities of medical staff are examples of difficulties that need a more thorough understanding of social obstacles. Concerns about data security are among the most serious problems with technology.

10.8   INDUSTRY 4.0 IN HEALTHCARE

A number of cutting-edge technologies are being used by the healthcare sector as part of Industry 4.0. These include AI, IoT, digitalization, machine learning, human psychology, augmented reality (AR), and big data mining [5]. These state-of-the-art technologies are designed to make users' lives easier by actively helping with illness identification and treatment. The industry was prepared to go forward with Industry 5.0, but there were a number of obstacles that necessitated reviewing this paper first. One of the most important things that modern medical research must take into account is the necessity for privacy and security. Healthcare must first understand the intricacies of privacy and security issues before they can be adequately prepared for Industry 5.0. When things, people, and data are linked, new insights may be uncovered in many settings [28]. Healthcare, education, development of natural resources, business and public sector organisation building, end-user computing, and other intelligent settings all get real-time solutions. When it comes to creating intelligent settings to address a wide range of real-world issues, there are three main advantages that stand out. Three of these factors are resource use, accessibility, and cost. Embracing Industry 4.0 will allow modern organisations to provide a multitude of real-time solutions for various organisational and commercial issues. Incorporating digital diagnostic systems into a wide range of digital projects has greatly improved the speed and accuracy of magnetic resonance imaging (MRI) and computed tomography (CT) scans. This made it much easier to provide prompt response based on a patient's medical history and past diagnoses.

10.8.1   THE ROLE OF IoMT IN HEALTHCARE

The authorization strategy has shifted from fixed desktop contexts to interactive cloud environments, thanks to IoT, which has resulted in massive amounts of data and services [25]. As a result, data-driven and organisational expenses go down. A healthcare provider's data and patient information may be quickly accessed in an emergency, allowing for faster treatment. Through IoT, data analysis and telemedicine have made it possible for people living in remote places to get high-quality healthcare. The healthcare business has long used an inclusive, egalitarian, and integrated paradigm to achieve transformation. Prioritising individual needs may enhance patient outcomes. In order to provide complete healthcare that is tailored to each individual's requirements, digital technology integration is essential. In order to contain the COVID-19 pandemic, mitigate its effects on people's lives, boost efficiency, make frontline workers safer, and lower death rates, the Internet of Medical Things (IoMT) has been adopted in conjunction with other techniques by a number of nations [29]. In terms of applications, technology, and security, the fast and global adoption of IoMT has been very beneficial. Secure IoMT apps based on his 4,444 research on technological security measures are now live. Improvements in practicality will also result from the ongoing research and development of new IoMT technologies that combine blockchain, artificial intelligence, and big data.

10.8.2 IoMT Architecture

The term "Internet of Medical Things" (IoMT) describes a network of networked sensors, medical equipment, and wearable tech that can gather, store, and analyse data. There are four distinct levels to the IoMT architecture, and they all cooperate to make healthcare data integration possible. There are four levels: perception, edge, network, and cloud computing. This section analyses each layer and its role in the IoMT structure.

10.8.2.1 Perceptual Layer

The perceptual layer is the initial layer of IoT architecture, responsible for acquiring data from various medical devices and sensors. This layer houses the user's physiological sensors, which include a glucose detector, heart rate monitor, and blood pressure monitor.

10.8.2.2 Edge Layer

The edge layer connects the perception and networking levels in an IoT architecture, making it the second layer of the design. This layer analyses data from sensors in real-time so that it may be sent via the Internet. Edge layers' design aims include scalability and the capacity to analyze large amounts of data.

10.8.2.3 Organizing Layer

The organising tier is the third and final tier of the IoMT design, and it can transport data acquired by the discernment layer to the cloud computing layer. This layer includes numerous wireless technologies and protocols used to share data across the internet, including Bluetooth, Zigbee, and Wi-Fi. The organising layer's speed, reliability, and security ensure that information transported from the edge layer to the cloud computing layer is neither lost nor corrupted.

10.8.2.4 Cloud Computing Layer

The cloud computing layer, which can store and analyze data collected from the recognition layer, is the final stage of IoMT architecture. This layer consists of several cloud computing services used for data storage and preparation, including Microsoft Sky Blue and Amazon Web Services (AWS). The cloud computing layer enables healthcare practitioners and analysts to access extensive data for developing novel cures and medications for diverse medical conditions.

10.9 IoMT DEVICES

To improve understanding of treatment and health outcomes, the Internet of Medical Things (IoMT) connects numerous healthcare technologies, frameworks, and equipment to the Internet. The spread of inexpensive and accessible IoMT as a result of IoT advances has freed up crucial administrative time in the healthcare sector. IoT devices may connect to the internet and collect real-time data for processing and analysis. This section looks into the characteristics of IoT devices that are well-suited to the healthcare web, such as their capabilities and potential improvements to healthcare delivery.

10.9.1 WEARABLE GADGETS

IoT refers to a broad spectrum of wearable devices that incorporate popular technology such as smartwatches, fitness monitors, and smart clothes. These gadgets may track a person's pulse, blood weight, sleep schedule, and their level of exercise. Connectivity between wearable devices and healthcare providers allows for continuous monitoring of patients' vitals. Wearable health trackers may alert healthcare providers to possible health concerns, for instance, if blood weight levels seen by the device exceed the average range.

10.9.2 SAVVY RESTORATIVE GADGETS

In order to provide restorative back, IoT based restorative gadgets, also known as clever therapeutic gadgets, are manufactured with a specific goal in viewpoint. They are able to communicate in real time with medical professionals and monitor vital signs including blood sugar, oxygen levels, and core temperature. When the health of patients is monitored in real-time, this occurs. To reduce the likelihood of both low and high blood sugar, an astute affront pump may regulate affront levels in accordance with glucose measurements.

10.9.3 TELEMEDICINE GADGETS

One IoT invention that could allow individuals to seek medical care without leaving their homes is telemedicine. Included in this category of devices are video conferencing tools, accessible testing frameworks, and portable health applications. Telemedicine has the potential to reduce the need for patients in remote or rural locations to physically see a healthcare provider, therefore improving access to healthcare services for such people. In certain cases, patients may save travel and lost work time by chatting with healthcare professionals remotely using video conferencing.

10.9.4 DOMESTIC GADGETS

IoT devices, sometimes called smart home gadgets, are integrated into houses to enhance convenience, safety, and comfort. Among these devices are locks, lights, and cunning indoor regulators. Healthcare delivery may also be radically altered by smart home devices that monitor patients' vitals and alert medical professionals to any unusual trends. For example, an intelligent bed may monitor sleep patterns and notify healthcare providers if there are any variations from standard sleep patterns, suggesting a potential health risk.

10.10 VARIOUS APPROACHES FOR SECURITY AND PRIVACY IN THE HEALTHCARE INDUSTRY

10.10.1 MUTUAL AUTHENTICATION AND ITS SOLUTION

Chen et al. [36] created a three-factor IoT-based protocol with the aim of resolving security problems. These problems are generally present in IoT systems for healthcare. The protocols that are used to guarantee privacy and control access are Key negotiation and authentication. The research also deliberates the use of biometrics to increase user experience along with protecting consumers' privacy. Asymmetric encryption and decryption are further used for security. To assess the security session key and protocol's soundness, integrity, and security are meticulously used. The research also compares the recommended procedure beside others of its kind to measure its performance. Efficiency, security, and computing cost are outperforms considering the proposed protocol.

10.10.2 SOLUTION FOR KEY AGREEMENT AND AUTHENTICATION

To enable optimal physical security, decentralization, data integrity, and accountability Yu et al. [37] proposed an architecture for security which is built on Blockchain and physically unclonable services. This protocol can alleviate several security concerns, comprising forgery attacks, impersonation, and session key leaks; it can also offer "anonymity," "mutual authentication," and "untraceability." To assess the scheme's security, testbed evaluations were executed by means of cryptographic primitives, simulation, and mathematical analysis.

10.10.3 A WAY OF LIGHTWEIGHT AUTHENTICATION

To protect the confidentiality and safety of comprehensive healthcare integration (CHI) in IoMT applications, RAPCHI (robust authentication protocol for IoMT-based CHI) was created by Kumar et al. [38]. Which provides patient-server-physician authentication and key agreement instead of keeping the information on the cloud. Also, it creates a session key which the doctor and patient might use during their session. The proposed methodology fulfils numerous security measures and is immune to a range of security risks. Moreover, on comparative analysis it was observed that RAPCHI outdoes alternative procedures in comparable circumstances—mainly in the event of a pandemic.

10.10.4 AN AUTHENTICATION SOLUTION FOR RFID SECURITY

In the field of smart healthcare, Wang et al. [39] studied the use of Radio Frequency Identification (RFID) technology and its effects on patient management, medical service labour cost reduction, and patient care improvement. It also highlights the dangers of misusing RFID tags, which might compromise patients' privacy and safety. To solve this problem, the authors present CRUSAP, an intelligent RFID authentication system that employs Bit-Crossing XOR rearrangement procedures in conjunction with a cloud server. To achieve its goals of maximising security and minimising resource usage, CRUSAP defends against attacks such as replay, desynchronization, forgery, and denial of service. Formal validation of the protocol has shown it to be safe and workable. The suggested protocol has proven to be capable of offering strong security at a reduced cost through security research and testing.

10.10.5 SOLUTION BASED ON ELLIPTIC CURVE CRYPTOGRAPHY

After looking into the protocol by Sahoo et al., Ryu et al. [40] found that it does not adequately secure patient confidentiality and is vulnerable to insider and privileged insider assaults. To solve these issues, they provide a new protocol that integrates biometrics and ECC to ensure secure communication in TMIS scenarios. By using BAN logic, the ROR model, and AVISPA, we evaluate the innovative protocol's resistance to many security vulnerabilities, including privileged insider attacks, stolen mobile device attacks, and insider attacks. Additionally, it protects the privacy of patients. With respect to processing expenses, communication costs, and security features, the new protocol outperforms the existing protocols and provides better protection. In their discussion of the evolution of information and communication technologies and IoT, Ahmed and Kannan [34] discuss its possible applications across several industries, including healthcare. Remote patient monitoring is one area where IoT and related wearables and sensors might enhance healthcare delivery. The MOTO 360 watch, a server, and an app for monitoring vital signs are the components of an encrypted and private RPM system. According to the review, the suggested approach might improve healthcare services and people's quality of life.

10.10.6 FEDERATED LEARNING APPROACH

Rahman et al. [41] introduced the idea of using the IoMT in health management and stressed the need of secure data management to keep personal information private. In order to solve the problems of insufficient training capacity and trust management, the authors suggest a hybrid federated learning architecture that uses blockchain smart contracts to manage trust and authenticate federated nodes. The solution provides end-to-end encryption and anonymity for IoMT data by use of DP. Evaluation of the system using deep learning applications on COVID-19 patients revealed promising results for the widespread and secure deployment of IoMT-based health management. Wang et al. [33] developed FRESH, a smart healthcare system that protects physiological data exchange against source inference attacks (SIAs) using federated learning (FL) and ring signature protection. Utilising edge computing devices, the system compiles physiological data gathered from wearables and processes it locally to train machine learning models. Next, the parameters of the model are sent to the main server in preparation for group training. The success rate of SIAs is decreased and the source of parameter adjustments is concealed by the ring signature. The batch verification approach that has been proposed also increases the efficiency of signature verification. A FRESH system would work well with large-scale, multi-user smart healthcare systems.

10.10.7 A Solution Based on Blockchain

Garg et al. [42] presented the BAKMP-IoMT, a protocol for secure authentication and key agreement in an IoMT scenario that is based on the blockchain. While IoMT connects healthcare systems to mobile health applications and devices, it also leaves patients' private health information vulnerable to tampering. With BAKMP-IoMT, patients may have access to their medical records stored on a blockchain that is overseen by servers in the cloud. This solution also ensures that the servers and implanted medical devices can maintain keys securely. The protocol's security was verified using the AVISPA tool, and it was shown to have lower communication and computational costs than current methods, all while offering higher functionality and security. The simulation's findings demonstrate how BAKMP-IoMT affects performance metrics. Singh et al. [30] provide a decentralized healthcare financing system that employs non-interactive zero-knowledge proof and blockchain technology to guarantee security and anonymity. Specifically designed for resource-constrained devices in Internet of Things networks, this technology reduces communication costs by utilizing lightweight cryptographic techniques as shown in Figure 10.2 The system may readily be extended to other financial systems, although its primary purpose is micro-level healthcare financing. It can also be audited, is very lightweight, and has transaction validation times of milliseconds.

Healthcare System: This refers to the different healthcare providers that work with patient data, such as physicians, hospitals, clinics, and other healthcare organisations.

Blockchain Network: A decentralised ledger that safely logs and archives transactions over a computer network. The blockchain creates a safe and unchangeable chain of data by include a cryptographic hash of each block.

Smart contracts: These are self-executing agreements that have the parties' stipulations explicitly encoded into the code. Smart contracts have the potential to automate several procedures in the healthcare industry, including authorization for data exchange, consent for data usage, and access control.

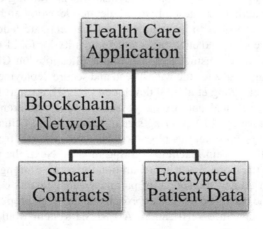

FIGURE 10.2 Healthcare application using blockchain technology.

Patient data is encrypted and securely stored on the blockchain network. Smart contracts and cryptographic keys are used to control the access to this data, ensuring that only authorised individuals or entities could view or interact with it.

10.10.8 A STRATEGY BUILT ON HOMOMORPHIC ENCRYPTION

To hindrances in implementing IoT in healthcare is discussed in the paper Othman et al. [35] with an emphasis on energy usage and security. When medical gadget are connected wirelessly used in transmitting medical data may raise security concerns and need a lot of energy. To address these concerns, the article offers EPPADA, which is a data aggregation system which employs homomorphic encryption used in protecting healthcare data and instantaneously individuals' privacy s protected. E-health sensor-shield platform was built experimentally and the research indicates that latency was reduced, computational cost, and communication overhead along with maintaining security.

10.10.9 AN APPROACH USING LIGHTWEIGHT CRYPTOGRAPHIC PRIMITIVES

Addressing security and privacy concerns a privacy-preserving mutual authentication strategy for IoT-based healthcare systems was proposed by Das et al. [32]. The proposed solution employs effective cryptographic techniques like as XOR, concatenation, and hashing. To protect and processing of the IoT devices by employing the technique by securely connecting permitted devices to a gateway which prevents unauthorized users from accessing the healthcare system. The proposed methodology outperforms in terms of security and speed.

10.10.10 A DATA-AGGREGATION APPROACH

Edge-enabled privacy-preserving data aggregation (EEPPDA) as a method for protecting medical information in IoT-based medical systems is proposed by Bhowmik et al. [31]. The data gathered from patients is encrypted by use of homomorphism. The data is transferred to cloud servers by pooling it on to the edge servers which allows licensed healthcare practitioners to verify, review, and manage data which is encrypted. The EEPPDA system guarantees patient privacy and medical information integrity, safeguards against potential risks as well checks data fidelity. Computing complexity and communication overhead is significantly reduced in comparison to existing techniques.

10.11 DIGITAL TECHNOLOGIES' EFFECTS ON THE HEALTHCARE INDUSTRY

Growing numbers of diseases and mounting pressure on the global healthcare system mean that person-centred care (PCC), which promotes patient engagement in care delivery and self-care, has to be given more attention. The number of chronic illnesses is rising as a result of demographic shifts [43]. PCC is, therefore, an essential component of chronic care.

Finding cutting-edge remedies to the pandemic is the responsibility of the entire society due to the present COVID-19-caused epidemic. Globally, the COVID-19 epidemic has prompted digital transformations in a number of industries, most notably the healthcare industry [44]. In reaction to the pandemic's initial stages, healthcare institutions swiftly embraced digital solutions and cutting-edge technologies.

The availability of health-related data has made it easier than ever for patients to get guidance at their convenience. One biosensor that can measure both heart rate and blood oxygen levels is the pulse oximeter. Serum haemoglobin saturation is an indicator of the respiratory system's ability to provide blood with oxygen [45]. The latest low-cost variants allow data transfer and measurement recording via conventional connectors. During the present pandemic, home monitoring of paucisymptomatic patients would be incredibly helpful in order to promptly respond in a state where saturation deteriorated. As a result, the launch of these digital gadgets and software coincides with the growing trend of on-demand healthcare that was noted during the COVID-19 outbreak in particular.

10.12   BIG DATA ENHANCING THE ADMINISTRATION OF HEALTHCARE

In the last several years, big data analytics (BDA) has increasingly affected healthcare businesses [28]. Clinical data is currently having a significant influence on healthcare organizations since it serves as the basis for a variety of digital solutions that are developed in this field. After ten years of progress in digitizing medical data, government and private businesses are working together to continue this age of open information in healthcare. Many public entities, including the federal government, have made data sets spanning decades available to the public in an effort to improve healthcare decision-making [46]. The sage of big data in healthcare is s shown in Figure 10.3. However, as more data becomes available to the public, maintaining patient privacy is becoming increasingly crucial. Ensuring appropriate security measures for the entities disclosing personal data is crucial.

Among clinical stakeholders, the idea of evidence-based medicine has grown in favour. Under this method, the best available scientific data is used to inform decisions about a particular therapy. Large data sets combine to generate big data algorithms, which are the greatest source of evidence for patient information when compared to individual patient data sets with smaller data sets. Reason studies indicate that organisations who scale up BDA more quickly should see favourable effects sooner. Other businesses are starting to explore analytics as a result since they don't

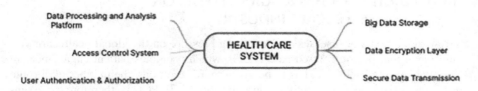

FIGURE 10.3   Bigdata in healthcare.

want to fall behind. The providers and BDA couples must always enhance healthcare values while maintaining or raising the calibre of services. There are several strategies to add value to this system, including making sure that care is affordable and getting rid of fraud, abuse, and waste.

10.13 INTEGRATING NUMEROUS COVID-19-SPECIFIC TOOLS WITH EHRs

Medical professionals rely on electronic health records (EHRs), which are digital dossiers detailing a patient's current and prior health status, to provide optimal treatment and health-related services. Computer systems that handle the acquisition, retrieval, storage, connection, and modification of multimedia data also hold these records. A wide variety of EHRs have been developed over the years, especially after the year 2000, when patients were held accountable for their own health and for gathering and preserving all of their medical information [45]. With the help of this effort, doctors may quickly ascertain the clinical characteristics of their patients and link any concerning physical state to current medical data. As a result, it permits the professions to change from personal experience-based medicine to evidence-based medicine. This facilitates better patient-physician contact and increases the accessibility of private medical records. Innovative personal illness monitoring devices, such as those that facilitate communication, medical picture processing, and acquisition, let doctors provide individualized care [10].

Several COVID-19-specific solutions for crisis management may be constructed using the EHR. The aforementioned solutions cover a wide range of topics, including secure communications, electronic check-ins, automatic triaging, standard ordering and documentation, and real-time data analytics. Medical imaging, socio-behavioural traits, and the patient's environment are all pieces of data that the EHR compiles. One advantage of these electronic devices is that they make it easier to access a patient's medical history as well as manage data that is pertinent to their current healthcare [10]. Because of the shorter time lag between tests, diagnosing and treating medical diseases is now timelier, which has a favourable effect on treatment speed.

Delivering prompt, high-quality medical care to patients is the ultimate aim of healthcare institutions. Immediate public health monitoring is made possible by digital technology, such as the EHR system, which gives pertinent information about individuals and eventually raises the standard of treatment. For the time being, we may overlook the complexity that big data adds in order to see how it can reduce confusion and delays in management. The effective application of cutting-edge digital technologies in the healthcare industry requires careful planning. Digital records, however, are essential to patients' survival.

10.14 MITIGATING DIGITALIZATION'S PRIVACY AND SECURITY RISKS IN THE HEALTHCARE INDUSTRY

Digital technology is advancing quickly in today's commercial world. Healthcare administration may find it difficult to stay up with the digital revolution. There could be setbacks with regard to patient and other healthcare stakeholders' security and privacy in between implementing and embracing new digital technology. Although management is crucial in many fields, it is particularly crucial in the medical field as patient safety is the most critical factor [47]. Medical device security is an important topic of continuous research across the globe, and as a consequence, standards are being established to regulate their production.

10.15 DANGERS FACING THE HEALTHCARE INDUSTRY

Data security is an area where the healthcare sector is lacking compared to other top organisations, making it an easy target for medical information thieves [13]. The rising tide of cybercrimes in this technologically advanced age makes it all the more urgent to devote resources to protecting patients' personal information and cutting-edge healthcare technology from unauthorised access, which is why this study zeroes in on these issues in the healthcare sector.

The development of medical technology has led to an increased integration of the healthcare system [48]. Better productivity, automated operations, lower error rates, and remote monitoring are just a few of the advantages of interconnectivity. The management of severe, long-term diseases is being revolutionised by these advantages. Medical professionals may now effectively monitor and adjust implanted devices remotely using interconnected technology, doing away with the need for in-hospital treatments altogether.

The healthcare business faces increased cyber dangers and is more susceptible to intrusions on a worldwide scale because of the inadequate security mechanisms. Over 81% of the 223 US companies reviewed had compromised patient information in 2015 alone, amounting to over 110 million records [49]. Cyberattacks have increased by 300% in the last three years, and only 50% of US healthcare workers believe they are secure [49]. Because of its informational weakness and its reputation as a vulnerable target, the healthcare industry is often targeted by cyberattacks.

Attacks on the healthcare system have up until now mostly been driven by political and financial gain. In the future, though, cyberattacks might be deliberate or accidental, and they might even alter test findings or blood types, causing harm. The potential for damaging cyberattacks on medical equipment is another worrying scenario. In 2014, over 300 medical gadgets were deemed potentially hazardous. Patients who use IoT technology and had poor experiences with cyber security may become reluctant to provide researchers and doctors with information, which might further obstruct treatment and lead to inadequate and inaccurate care.

10.16 CREATING PROGRAMS FOR THE ADMINISTRATION
OF MEDICAL EQUIPMENT

The main danger in identifying cyberattacks is that many go unreported or unnoticed. Hospital biomedical engineering teams create and oversee medical equipment management programs (MEMPs) [47]. When it comes to privacy and security breaches caused by attacks on digital devices, MEMPs protect the reliability and security of medical equipment. Medical equipment is increasingly integrated with data and communication technologies, keeping pace with the fast progress in technology. However, because such divergent technologies are interconnected, there is a steady rise in both internal and external security concerns. Therefore, it is necessary to identify and prioritize essential systems in order to reduce these hazards from medical equipment. Although several MEMPs have been created and implemented in healthcare organisations for more than 30 years, there has been a dearth of research providing an ideal concept that considers affordability, safety, and reliability when it comes to the service delivery of these medical equipment devices [50].

10.17   HOW ARE MEDICAL RECORDS KEPT IN EHR SOFTWARE?

EHRs have the ability to improve patient care by increasing the accessibility of health-related data, which is in line with the previous points. Cybersecurity, on the other hand, is concerned with protecting computer systems and the data they hold from infiltration and inadvertent or detrimental disturbance [51]. The absence of medical information integrity and confidentiality has raised worries about the inadequate cybersecurity in the healthcare industry. EHRs are physical items that are dispersed over several regions. Because of this, unauthorized people can access the platform and devices, corrupt all of the data on them, or even program the device and take out the security keys. The purpose of this study is to bring attention to the security and privacy risks connected with digital technologies in the healthcare business. This is important since the industry must undergo significant development before it can fully use new technology to provide patients with high-quality treatment.

10.18   FUTURE SCOPE

10.18.1   FURTHER RESEARCH OF SECURITY AND PRIVACY REGULATIONS IN HEALTHCARE

The healthcare sector is legally mandated to deploy security and privacy safeguards to protect the personal data of patients. Encryption technology protects electronic data from illegal access. Access restrictions limit access to patient data depending on roles and authorisation levels. Medical practitioners are subject to restrictions like insurance coverage. HIPAA and GDPR mandate strict privacy and security measures for patient information. The healthcare business utilizes both technology and policy-based measures to ensure patient data security and privacy. Further studies might focus on the healthcare sector's privacy and confidentiality rules.

10.18.2   EXPLORATION INTO HOW DIGITALIZATION INFLUENCES PATIENT OUTCOMES

In the sector of healthcare, digitalization has had a substantial impact on patient outcomes. EHRs are one instrument used to measure this impact. Increased access to a patient's medical history through digitalization can upturn diagnosis and treatment planning. One of the additional strategies is using telemedicine, which allows patients to consult with medical specialists distantly. Patients from remote locations who cannot travel for in-person appointments will get more access to treatment. Technologies like machine learning and big data have influenced in examination of vast amounts of patient data which enables in identifying patterns.

10.18.3 EXPLORATION OF SECURITY AND PRIVACY CONCERNS OF WEARABLE TECHNOLOGY

There are numerous methods to evaluate wearable devices' security and privacy risks. A thorough analysis of the hardware and software of the device is necessary to identify any susceptibilities that hackers could abuse. One of the techniques is to use penetration testing, where the system finds the security flaws from the actual attacks on the device. A combination of technological research, testing, and user feedback is required to identify and resolve privacy and security concerns with wearable technology.

10.18.4 ASSESSMENT OF ARTIFICIAL INTELLIGENCE'S ROLE IN HEALTHCARE

Natural language processing (NLP) is one of the areas where it allows the medical professionals to input patient data and receive computerized recommendations for diagnosis or treatment. Easy access to medical information is provided by the usage of AI-driven chatbots and virtual assistants, which helps in expedite better patient-provider services. Moreover, to examine medical images and spot possible health problems doctors are using diagnostic and imaging analytic technologies powered by AI. AI technologies have the power to improve patient healthcare monitoring and enable the best possible way to deliver the healthcare in the modern world.

10.18.5 AN ANALYSIS OF BLOCKCHAIN TECHNOLOGY'S POSSIBLE APPLICATIONS IN HEALTHCARE

Blockchain technology is one of the latest technologies which is revolutionize the healthcare industry. This technology offers safe, decentralized, and open methods for exchanging and storing patient data. Patient data can be easily efficiently and sagely can be exchanged by the medical professional by using blockchain platforms. Smart contract is another approach that strength patients and insurance companies to save time and money. Supply chain management industry also can use blockchain technology which ensures that the validity and traceability medical equipment and medications. Overall, the increased security and efficiency that blockchain technology may deliver to the healthcare industry will benefit patients and healthcare providers.

10.18.6 EXAMINING THE EFFECT OF PATIENT ENGAGEMENT ON SECURITY AND PRIVACY

Patient engagement is critical to the healthcare industry's security and privacy. When patients are actively involved in the process of protecting their own personal information and medical records, healthcare providers may ensure that patients understand their responsibilities and rights in this area. This entails teaching patients the value of protecting their personal data and giving them the instruments and resources necessary to do so.

10.19   CONCLUSION

Healthcare regulations, methods, and activities will need to be modernised to keep up with digitalization and meet current employment demands. Evaluating healthcare personnel's proficiency with digital technologies is necessary for addressing concerns about patient safety and incorporating digital technologies into the workplace. Digital health systems and other health information technologies are increasingly recognized as potential replacements for the fragmented, antiquated paper-based health record systems in developing economies.

Consequently, there has been a rise in the requirement for healthcare establishments to integrate diverse digital technology to optimise their processes. The availability of more dependable, affordable, and low-power hardware and software, increased internet usage, and the emergence of several widely reported initiatives across most countries are some of the main causes of the sharp increase in the use of these technologies. With healthcare services becoming ever more time-sensitive, it is anticipated that these digital improvements will lead to substantial cost savings and improved efficiency in the healthcare industry. There is hope that digital health systems may improve healthcare in countries that are economically still developing.

Although the opportunities and benefits that digitalization brings to the healthcare industry, context-specific challenges sometimes impede its successful adoption.

Healthcare firms should prioritise the social climate of the workplace and cultivate a positive attitude if they want to improve their response to digitalization. The implementation of new technologies needs cooperative, organised, and pragmatic support. Although healthcare digitalization offers huge potential advantages, several experts have cautioned that not all computerised healthcare systems meet expectations. The anticipated improvements brought about by the digitalization of healthcare may not always materialise as expected, according to researchers.

REFERENCES

[1] K.Y. Chau, M.H.S. Lam, M.L. Cheung, E.K.H. Tso, S.W. Flint, D.R. Broom, G. Tse, K.Y. Lee, Smart technology for healthcare: Exploring the antecedents of adoption intention of healthcare wearable technology, Health Psychol. Res. 7 (1) (2019).

[2] P. Tyrvainen, T. Kilpelainen, M. Jarvenpaa, Patterns and measures of digitalisation in business unit communication, Int. J. Bus. Inf. Syst. 1 (1–2) (2005) 199–219.

[3] J. Manyika, S. Lund, J. Bughin, Digital globalization: The new era global flows, Tech. rep., McKinsey Global Institute, 2016.

[4] A. Bereznoy, Multinational business in the era of global digital revolution, Mirovaia Ekon. Mezhdunarodnye Otnos. 62 (9) (2018) 5–17.

[5] V.V. Popov, E.V. Kudryavtseva, N. Kumar Katiyar, A. Shishkin, S.I. Stepanov, S. Goel, Industry 4.0 and digitalisation in healthcare, Materials 15 (6) (2022) 2140.

[6] D.K. Sharma, D.S. Chakravarthi, A.A. Shaikh, A.A.A. Ahmed, S. Jaiswal, M. Naved, The aspect of vast data management problem in the healthcare sector and implementation of cloud computing technique, Mater. Today: 80: Proc. (2021) 3805–3810. https://doi.org/10.1016/j.matpr.2021.07.388

[7] F. Jiang, Y. Jiang, H. Zhi, Y. Dong, H. Li, S. Ma, Y. Wang, Q. Dong, H. Shen, Y. Wang, Artificial intelligence in healthcare: Past, present and future, Stroke Vasc. Neurol. 2 (4) (2017).

[8] Digital health—statistics & facts, 2021, Online; accessed 7-October-2022, www.statista.com/topics/2409/digital-health/.

[9] M.R. Cowie, J.I. Blomster, L.H. Curtis, S. Duclaux, I. Ford, F. Fritz, S. Goldman, S. Janmohamed, J. Kreuzer, M. Leenay, et al., Electronic health records to facilitate clinical research, Clin. Res. Cardiol. 106 (1) (2017) 1–9.

[10] S.P. Dash, The impact of IoT in healthcare: Global technological change & the roadmap to a networked architecture in India, J. Indian Inst. Sci. 100 (4) (2020) 773–785.

[11] K. Abouelmehdi, A. Beni-Hssane, H. Khaloufi, M. Saadi, Big data security and privacy in healthcare: A review, Procedia Comput. Sci. 113 (2017) 73–80.

[12] H. Landi, Average cost of healthcare data breach rises to 7.1 M, according to IBM report, Fierce Health. July 29 (2020).

[13] C.S. Kruse, B. Frederick, T. Jacobson, D.K. Monticone, Cybersecurity in healthcare: A systematic review of modern threats and trends, Technol. Health Care 25 (1) (2017) 1–10.

[14] N.S. Abouzakhar, A. Jones, O. Angelopoulou, Internet of things security: A review of risks and threats to healthcare sector, in: 2017 IEEE International Conference on Internet of Things (IThings) and IEEE Green Computing and Communications (GreenCom) and IEEE Cyber, Physical and Social Computing (CPSCom) and IEEE Smart Data (SmartData), IEEE, 2017, pp. 373–378.

[15] I. FBI, Internet crime report 2021, 2021.

[16] M. Ventura, C.M. Coeli, Beyond privacy: The right to health information, personal data protection, and governance, Cadernos de Saude Publica 34 (2018).

[17] E.M. Kwiatkowska, M. Skórzewska-Amberg, Digitalisation of healthcare and the problem of digital exclusion, Cent. Eur. Manag. J. 27 (2019) 48–63.

[18] N. Chouliaras, G. Kittes, I. Kantzavelou, L. Maglaras, G. Pantziou, M.A. Ferrag, Cyber ranges and testbeds for education, training, and research, Appl. Sci. 11 (4) (2021) 1809.

[19] M. Evans, Y. He, L. Maglaras, I. Yevseyeva, H. Janicke, Evaluating information security core human error causes (IS-CHEC) technique in public sector and comparison with the private sector, Int. J. Med. Inform. 127 (2019) 109–119.

[20] M. Herrmann, P. Boehme, T. Mondritzki, J.P. Ehlers, S. Kavadias, H. Truebel, et al., Digital transformation and disruption of the health care sector: Internet-based observational study, J. Med. Internet Res. 20 (3) (2018) e9498.

[21] J. Roski, G.W. Bo-Linn, T.A. Andrews, Creating value in health care through big data: Opportunities and policy implications, Health Aff. 33 (7) (2014) 1115–1122.

[22] B. Chae, Mapping the evolution of digital business research: A bibliometric review, Sustainability 14 (12) (2022) 6990.

[23] A. Lipsmeier, A. Kühn, R. Joppen, R. Dumitrescu, Process for the development of a digital strategy, Proc. CIRP 88 (2020) 173–178.

[24] P. Barwise, L. Watkins, The evolution of digital dominance, in: Digital Dominance: The Power of Google, Amazon, Facebook, and Apple, Oxford University Press, 2018, pp. 21–49.

[25] C.A. Tavera Romero, J.H. Ortiz, O.I. Khalaf, A. Ríos Prado, Business intelligence: Business evolution after industry 4.0, Sustainability 13 (18) (2021) 10026.

[26] B. Richter, J.H. Hanf, Cooperatives in the wine industry: Sustainable management practices and digitalisation, Sustainability 13 (10) (2021) 5543.

[27] M. Alloghani, D. Al-Jumeily, A. Hussain, A.J. Aljaaf, J. Mustafina, E. Petrov, Healthcare services innovations based on the state of the art technology trend industry 4.0, in: 2018 11th International Conference on Developments in ESystems Engineering (DeSE), IEEE, 2018, pp. 64–70.

[28] G. Manogaran, C. Thota, D. Lopez, R. Sundarasekar, Big data security intelligence for healthcare industry 4.0, in: Cybersecurity for Industry 4.0, Springer, 2017, pp. 103–126.

[29] A.H.M. Aman, W.H. Hassan, S. Sameen, Z.S. Attarbashi, M. Alizadeh, L.A. Latiff, Iomt amid COVID-19 pandemic: Application, architecture, technology, and security, J. Netw. Comput. Appl. 174 (2021) 102886.

[30] R. Singh, A.D. Dwivedi, G. Srivastava, P. Chatterjee, J.C.-W. Lin, A privacy preserving internet of things smart healthcare financial system, IEEE Internet Things J. (2023) 18452–18460.

[31] T. Bhowmik, I. Banerjee, EEPPDA—Edge-enabled efficient privacy-preserving data aggregation in smart healthcare internet of things network, Int. J. Netw. Manag. (2023) e2216.

[32] S. Das, S. Namasudra, Lightweight and efficient privacy-preserving mutual authentication scheme to secure internet of things-based smart healthcare, Trans. Emerg. Telecommun. Technol. (2023) e4716.

[33] W. Wang, X. Li, X. Qiu, X. Zhang, J. Zhao, V. Brusic, A privacy preserving framework for federated learning in smart healthcare systems, Inf. Process. Manage. 60 (1) (2023) 103167.

[34] M.I. Ahmed, G. Kannan, Secure and lightweight privacy preserving internet of things integration for remote patient monitoring, J. King Saud Univ.-Comput. Inf. Sci. 34 (9) (2022) 6895–6908.

[35] S.B. Othman, F.A. Almalki, C. Chakraborty, H. Sakli, Privacy preserving aware data aggregation for IoT-based healthcare with green computing technologies, Comput. Electr. Eng. 101 (2022) 108025.

[36] C.-M. Chen, Z. Chen, S. Kumari, M.-C. Lin, LAP-IoHT: A lightweight authentication protocol for the internet of health things, Sensors 22 (14) (2022) 5401.

[37] S. Yu, Y. Park, A robust authentication protocol for wireless medical sensor networks using blockchain and physically unclonable functions, IEEE Internet Things J. 9 (20) (2022) 20214–20228.

[38] V. Kumar, M.S. Mahmoud, A. Alkhayyat, J. Srinivas, M. Ahmad, A. Kumari, RAPCHI: Robust authentication protocol for IoMTbased cloud-healthcare infrastructure, J. Supercomput. 78 (14) (2022) 16167–16196.

[39] X. Wang, K. Fan, K. Yang, X. Cheng, Q. Dong, H. Li, Y. Yang, A new RFID ultra-lightweight authentication protocol for medical privacy protection in smart living, Comput. Commun. 186 (2022) 121–132.

[40] J. Ryu, J. Oh, D. Kwon, S. Son, J. Lee, Y. Park, Y. Park, Secure ECCbased three-factor mutual authentication protocol for telecare medical information system, IEEE Access 10 (2022) 11511–11526.

[41] M.A. Rahman, M.S. Hossain, M.S. Islam, N.A. Alrajeh, G. Muhammad, Secure and provenance enhanced internet of health things framework: A blockchain managed federated learning approach, IEEE Access 8 (2020) 205071–205087.

[42] N. Garg, M. Wazid, A.K. Das, D.P. Singh, J.J. Rodrigues, Y. Park, BAKMP-IoMT: Design of blockchain enabled authenticated key management protocol for internet of medical things deployment, IEEE Access 8 (2020) 95956–95977.

[43] E. Granström, C. Wannheden, M. Brommels, H. Hvitfeldt, M.E. Nyström, Digital tools as promoters for person-centered care practices in chronic care? Healthcare professionals' experiences from rheumatology care, BMC Health Serv. Res. 20 (1) (2020) 1–15.

[44] D. Golinelli, E. Boetto, G. Carullo, A.G. Nuzzolese, M.P. Landini, M.P. Fantini, et al., Adoption of digital technologies in health care during the COVID-19 pandemic: Systematic review of early scientific literature, J. Med. Internet Res. 22 (11) (2020) e22280.

[45] F. Girardi, G. De Gennaro, L. Colizzi, N. Convertini, Improving the healthcare effectiveness: The possible role of EHR, IoMT and blockchain, Electronics 9 (6) (2020) 884.

[46] P. Groves, B. Kayyali, D. Knott, S. V. Kuiken, The 'big data' revolution in healthcare: Accelerating value and innovation (2016). https://www.mckinsey.com/industries/healthcare/our-insights/the-big-data-revolution-in-us-health-care

[47] D.-W. Kim, J.-Y. Choi, K.-H. Han, Medical device safety management using cybersecurity risk analysis, IEEE Access 8 (2020) 115370–115382.

[48] L. Coventry, D. Branley, Cybersecurity in healthcare: A narrative review of trends, threats and ways forward, Maturitas 113 (2018) 48–52.

[49] G. Martin, P. Martin, C. Hankin, A. Darzi, J. Kinross, Cybersecurity and healthcare: How safe are we? BMJ 358 (2017).

[50] H. Mahfoud, A. El Barkany, A. El Biyaali, Preventive maintenance optimization in healthcare domain: Status of research and perspective, J. Qual. Reliab. Eng. 2016 (2016).

[51] Gupta, M., Ahmed, S., Kumar, R., Altrjman, C. (Eds.), Computational Intelligence in Healthcare: Applications, Challenges, and Management, CRC Press, 2023.

11 AI and ML Fundamentals
A primer for healthcare professionals

Somya Srivastava, Gaurav Dubey, Disha Mohini
Pathak, Kamlesh Sharma, and Kuldeep Kumar

11.1 INTRODUCTION

Many new fields are being introduced by advancements in artificial intelligence (AI) and machine learning (ML). Machine intelligence, often known as AI, is the ability of a machine to perform tasks that a human brain can. Because AI allows robots to think, reason, grasp languages, and learn, it has revolutionized human interaction with technology [1]. Conversely, machine learning is a subfield of AI that deals with the use of algorithms to acquire new knowledge by analyzing existing data, drawing conclusions or predictions based on those conclusions, and then applying those conclusions or forecasts in the actual world. They have a profound impact on both their industries and their environment.

As shown in Table 11.1 ([2]), AI has radically boosted healthcare. The potential for disruptive innovations in healthcare is discernible from the escalating pace of expansion in the sector coupled with extensive financing by many startups. Measurable improvements in diagnosis accuracy rates and outcomes among patients together with high adoption rates of AI technologies in hospitals are indicators of the tangible contributions made by AI towards increasing effectiveness in healthcare delivery process. Not only does the value of these two technologies extend beyond technological advancement per se, but also represents changes in problem solving approaches as well as operational performance enhancement practices. AI and ML are seen as potential methods for tapping useful insights from expansive data sets today because more information is being produced and consumed second by second than ever before [3]. Artificial intelligence systems have led to some of the most significant breakthroughs in the healthcare industry, particularly in the areas of diagnosis, treatment planning, and patient monitoring. This therefore makes it a simple matter to improve customer relations using tailored suggestions.

The utilization levels of AI tools in the healthcare sector have experienced a significant and rapid increase, both in terms of depth and scope, as indicated in Table 11.2. This suggests the widespread adoption of technological advancements and the expansion of their application in various areas. Hence it is anticipated that utilizing such tools will comprise major activities in delivery of healthcare to mankind

DOI: 10.1201/9781003473435-11

TABLE 11.1
AI in Healthcare: Key Statistics

Statistic	Value	Year	Commentary
Global AI in Healthcare Market Size	USD 6 billion	2021	Initial valuation marking the rapid growth phase
Projected Market Size	USD 45 billion	2026	Anticipated growth reflecting increasing adoption
Compound Annual Growth Rate (CAGR)	40%	2021–2026	Demonstrates the sector's dynamic expansion
Investment in AI Healthcare Startups	USD 4 billion	2022	High investment levels indicating confidence in the sector
Number of AI Healthcare Startups Worldwide	Over 1,000	2022	Shows the burgeoning entrepreneurial interest in AI healthcare
Adoption Rate of AI Technologies in Hospitals	35%	2022	Indicates a growing integration of AI in clinical settings
Increase in Diagnostic Accuracy Through AI	20% Improvement	2022	Highlights AI's impact on enhancing diagnostic processes
Reduction in Hospital Readmission Rates	15% Reduction	2022	AI's role in improving patient outcomes and care continuity
Patient Engagement Through AI Platforms	50% Increase	2022	Enhanced interaction and monitoring via AI-driven tools

including, but not limited to, early diagnostic imaging support, individualized therapy, and aid during global health emergencies [4]. The healthcare business has experienced a profound and disruptive revolution due to the use of AI and ML. This technological advancement has had a huge influence on an industry that is highly complex and delivers essential services but has historically struggled with inefficiency. They have made advances such as medical image analysis using computer algorithms possible since they are more accurate in most cases when compared to human experts and radiologists who may miss important aspects due to their complex nature. Algorithms are being trained to predict patients' future encounters by integrating patterns from patient health data thereby enabling pre-emptive medical action. AI and ML are also driving towards new ways of working within genetic medicine that will result in operational efficiencies around practices such as drug discovery.

Investment trends and their relative importance in the AI healthcare industry are considered in Figure 11.1. With medical data becoming more complex by the day and healthcare delivery needing to be thus more efficient, AI and ML are vital. These can be utilized as tools for dealing with large amounts of data better as well as analyzing them to generate more accurate diagnoses, treatment plans, and patient outcomes. Nevertheless, widespread deployment of these systems in healthcare does not come without challenges; there should be a sufficient allocation of resources toward safeguarding data privacy, as well as addressing ethical concerns related to the use of

TABLE 11.2
Dynamic and Rapid Progression of AI in Healthcare

Year	Milestone/Development	Impact/Significance
2010	Initial applications of AI in diagnostic imaging	Marked the early use of AI for enhancing the accuracy and efficiency of diagnostics
2011	IBM Watson's success on Jeopardy!	Highlighted AI's potential for processing and analysing large datasets
2012	The emergence of deep learning techniques	Improved the capabilities of AI in image recognition, leading to better diagnostic tools
2014	Google DeepMind's advancements in AI	Paved the way for further research and development in AI for health data analysis
2016	FDA approvals for AI-based diagnostic tools begin	Signified regulatory recognition and the increasing trust in AI healthcare solutions
2017	Introduction of AI in wearable health devices	Expanded AI's role in continuous patient monitoring and personal health management
2018	Growth of AI-driven personalized medicine	Leveraged AI for analysing genetic information for personalized treatment plans
2019	Expansion of AI applications in EHR systems	Enhanced patient data management, predictive analytics, and operational efficiency
2020	AI for COVID-19 pandemic response	Utilized in diagnostics, drug discovery, and epidemiological forecasting
2021	Increase in telehealth and virtual care AI applications	Facilitated remote patient monitoring, triage, and treatment amid the pandemic
2022	Advancements in AI-powered robotic surgery	Improved surgical precision, reduced human error, and minimized recovery times
2023	Expansion of AI in mental health applications	Provided support for mental health assessments, therapy, and monitoring

AI/ML in medicine. Additionally, efforts should be made to minimize algorithmic bias and integrate these technologies into existing health systems as needed.

A comprehensive understanding of technology is essential for anyone aspiring to be a leader or pioneer in the healthcare industry, as the introduction of ML and AI has revolutionized the field. This chapter seeks to provide an organized presentation on AI and ML concepts relevant only to medical professionals. It aims to elucidate what AI and ML really mean and how they can be applied within healthcare, their impact on medicine, and the challenges they pose.

11.2   MACHINE LEARNING MODELS IN HEALTHCARE

Before we get into AI or ML, it is important to first understand some key principles and terms behind these game-changing technologies. This section introduces basic concepts behind AI as well as how we can forecast analysis on big data through them simulating human cognition using ML.

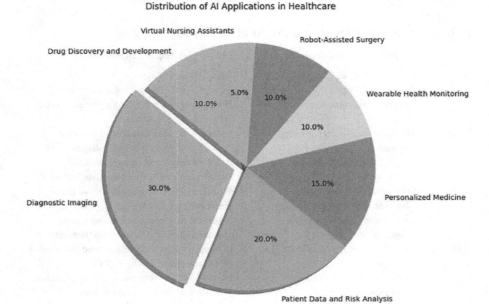

FIGURE 11.1   Relative importance and investment focus areas.

TABLE 11.3
Emerging Healthcare Industries

Key AI Player	AI Platform/Product	Services for Healthcare
IBM Watson	Watson Health	AI-powered analytics for diagnosis, treatment recommendations, drug discovery
Google DeepMind	DeepMind Health	AI research for medical image analysis, health data analysis
Nvidia	Clara	AI and GPU-powered analytics for medical imaging and genomics
Microsoft	Project InnerEye	AI tools for medical image analysis, particularly in oncology
Siemens Healthineers	AI-Rad Companion	AI-based applications for medical imaging analysis
GE Healthcare	Edison Platform	AI-powered applications for imaging, operational efficiency, patient monitoring
Philips Healthcare	IntelliSpace AI Workspace	Integration of AI applications for imaging workflows and analysis

Artificial Intelligence (AI) is a broad term in computer science that involves creating intelligent machines which can perform tasks requiring human intelligence. These systems are meant for natural language processing alone and therefore include speech recognition systems, machine learning systems or strategic planning systems designed to solve complex problems among others [5]. The idea is not just to imitate but also enhance upon natural cognitive abilities demonstrated by people so they can do harder things better than before. In healthcare, AI involves developing various applications that mimic mental processes such as learning new information and solving problems through which data can be analyzed, patterns identified, educated decisions made and useful insights offered. There are two main categories of artificial intelligence systems used in medicine, namely, weak (narrow) and strong (general) [6].

Narrow AI, often known as weak AI is programmed with specific instructions on what task it should perform and how to do it; therefore, they have limited scope within which their operations can be executed, but this doesn't mean that such systems lack awareness about themselves or any other form of true intelligence. Examples of this kind include virtual personal assistants (e.g., Siri), recommendation algorithms used by shopping websites (e.g., Netflix), etc.

General AI not only understands what needs knowing so as to accomplish given objectives but also knows why these facts are necessary for achieving them just like humans do. Such machines possess ability learn from one domain then apply knowledge gained across different domains thereby showing some level artificial consciousness which most times remains theoretical due absence actual implementations capable reaching such high levels.

There are many ways that artificial intelligence mimics human thought: rule-based systems, for example, which use logic to derive conclusions from data; and machine learning, which involves algorithms that can directly gather patterns and insights from data.

Machine learning is a subset of AI that allows computers to gain knowledge from information and improve their performance without being explicitly programmed. The core of machine learning is the creation of autonomous algorithms that are able to access data and learn from it. This has greatly contributed to the advancement of AI in activities such as big data analysis.

There are three main types of ML:

11.2.1 Supervised Learning

Unsupervised learning requires training an algorithm with a labeled dataset where each sample has corresponding labels linking input data with desired output. In such cases, the algorithm learns how to make predictions or evaluations using input data while receiving feedback whenever these predictions do not match expected outcomes. Supervised learning is often used when historical data can be used to predict future behavior like detecting fraud [7]. For instance, one can train a model on patient pictures labelled either "with disease" or "without disease" so as to correctly classify new images. Different models for supervised learning are discussed in the following sections.

11.2.1.1 Logistic Regression

When there is one or more independent variable in the dataset, this can be investigated with logistic regression (LR). The binary output variable only has two results to measure the outcome. In medical field, LR models are used for finding relationships between a binary dependent variable such as presence/absence of disease and some independent variable like age, weight, blood pressure genetic markers, lifestyle choices etc. Medical decision-making is greatly improved by logistic regression, which gives probability on classifying a particular input point.

LR is a statistical model that utilizes the sigmoid function (logistic function) given by equation 11.1. The sigmoid function maps any real integer to a value within the range zero to one inclusive. This makes it possible for linear regression results to be transformed into probabilities.

$$p(x) = \frac{1}{1 + e^{-(\beta 0 + \beta 1 x 1 + \beta 2 x 2 \ldots \beta n x n)}} \tag{11.1}$$

Where $\beta 0$, $\beta 1$ ... βn are the coefficients of the model and x1, x2 ... xn represent the independent variables.

LR has been commonly used in predicting if someone will get some diseases [7]. For instance, LR models may be employed to determine the risk of heart disease among patients based on cholesterol levels, body mass index (BMI), age, and smoking status while evaluating diagnostic tests. It also helps in calculating probabilities about malignancy from imaging data together with biopsy results and patient information using LR yet still able to evaluate therapeutic approaches efficacy where we compare treatments outcomes for patients according to their demographic characteristics thereby finding out what works best for whom when where and why.

11.2.1.2 Decision Trees (DT)

A DT is a type of graph in which characteristics are tested using internal nodes, test results are shown by branches, and class labels are chosen after all attributes have been evaluated using leaf nodes. The paths that run from the base to the end of the tree symbolize classification rules. A decision tree typically begins with a patient's symptom or condition as the root. The tree uses the input data to branch out to different nodes, each posing a question about the patient's data, until a diagnosis or outcome prediction is made at the leaf nodes. To diagnose a patient with or without diabetes, a decision tree algorithm may inquire about the patient's glucose levels, BMI, age, and other factors to accurately categorize them as diabetic or non-diabetic.

11.2.1.3 Random Forest (RF)

In most cases, the "bagging" method is used to train the decision trees that make up an RF. Building many decision trees and merging them into one more accurate and consistent forecast is the basic idea. The prediction capabilities of an RF model are based on an ensemble of decision trees. The final prediction is obtained by integrating the predictions of all the trees, typically through majority voting in classification tasks, and each tree uses a randomly selected subset of patient data attributes to get its judgment. Every individual tree inside the forest has the ability to examine several subsets of data, including past admissions, medication compliance, diagnostic outcomes, and socioeconomic factors, in order to predict the likelihood of patient readmission, as an example. The RF algorithm utilizes the predictions of individual decision trees to assess whether a patient has a significant likelihood of being readmitted.

11.2.2 UNSUPERVISED LEARNING

Unsupervised learning is a type of machine learning algorithm that analyzes and processes data without any prior information about how the computer should assess it. It is intended for computers to comprehend the content, discover patterns, and recognize relationships in any given dataset. This method is commonly employed in clustering or dimension reduction tasks; it helps one segment customers for marketing, simplify complicated situations, among others [8].

In healthcare this would involve finding clusters of patients exhibiting related symptoms or having same illnesses which have not been grouped before. This exposes new variations or subtypes of illnesses.

11.2.2.1 K-Means Clustering

To group patients into separate entities, AI has allowed K-means methods to be employed. It is possible to pinpoint populations in need of specific interventions with segmentation; these include individualized healthcare as well as related strategies. Individuals who are diagnosed with different signs/symptoms could have cancer patients among them as well diabetics when analyzed based on clinical attributes and genetic data so that—other than these two related conditions-others can also be recognized. As such, this is essential for designing cancer personalized therapies and understanding etiology. In addition, for instance, service improvement and K-means clustering are tools that can be used in resource optimization at hospitals; through identifying patterns of patient visits different clusters enable healthcare organizations to schedule staff accordingly to facilitate management care efficiently in a hospital setting. This may help to identify patients who are at risk of contracting various diseases by analyzing their medical records in relation to K-Means Coincidentally, these patients should be placed in different layers so that measures are taken to avoid any potential illnesses thereby promoting healthy living among them.

11.2.2.2 Principal Component Analysis (PCA)

PCA starts off by identifying the primary component that explains the most variability in the set of data points; following that it goes on to compute a 2nd principal component that is orthogonal to the initial one but with the highest possible value when compared with all available alternatives. Each subsequent step is characterized by a repetition of this process until we gather adequate facts concerning the data structure by examining separate factors through their dimensions.

PCA is a statistical technique that uses an orthogonal transformation to identify linearly uncorrelated variables from a group of potentially related observations, known as the major components, after they have been extracted. It is widely used in healthcare for dimension reduction, data visualization, feature extraction etc. This approach provides a way of dealing with complex medical datasets.

PCA implements the following steps:

1. **Standardization:** Data standardization involves normalizing the data to ensure equal contribution of each variable in the analysis.
2. **Covariance Matrix Computation:** Calculating the covariance matrix to analyze the relationships between variables.
3. **Eigenvalue and Eigenvector Calculation:** Calculating eigenvalues and eigenvectors to determine the directions that optimize the variance (principal components).
4. **Dimensionality Reduction:** Dimensionality Reduction involves choosing a subset of major components that effectively account for the variance in the data.

11.2.3 Reinforcement Learning

This type of learning uses a system of rewards and penalties to compel the computer to solve a problem by itself. Human intervention is minimal or absent. It is often used in robotics, gaming, and navigation applications. It has applications in treatment optimization and robotic surgery, where the system learns to perform tasks with precision and minimal invasiveness.

11.2.3.1 Q-Learning

Q-learning is a reinforcement learning algorithm that is model-free that determines the value of actions, representing the "Q" in Q-learning, to show the effectiveness of a specific action in a certain state. The process involves assigning a Q-value to each action in every state, which indicates the anticipated benefit of choosing that action and then adhering to the best strategy. The algorithm enhances its Q-values by applying the Bellman equation, which progressively refines the Q-values by considering the reward obtained from an action and the highest anticipated future rewards. The goal is to acquire the optimal policy that maximizes the cumulative reward over a period.

11.2.3.2 Deep Q-Networks (DQN)

The system of Q-learning is enhanced by deep Q-networks using deep neural networks in approximating the Q-value function. This method makes it possible to handle environments having high-dimensional state space which is common in healthcare applications.

DQN refers to a reinforcement learning algorithm that incorporates Q-learning along with a deep neural network. Each available action has its corresponding Q-values produced by the neural network taking the state as input. Experience replay is utilized by DQN algorithm together with fixed Q-targets for purposes of improving stability during learning process. In this case, experience replay implies storing past agent's experiences and selecting small batches randomly for updating the network thereby reducing correlations in sequence of observations. Fixed Q-targets on other hand employ separate networks to generate Q-value targets during updates but are refreshed periodically using weights from training network.

11.2.4 DEEP LEARNING

Deep learning is an advanced field of machine learning that utilizes algorithms based on the structure and operation of neural networks in the brain. Artificial neural networks consist of multiple layers that have the ability to learn from large amounts of data, similar to human cognition [8]. Deep learning has significantly advanced the capabilities of AI, especially in challenging areas such as image processing, natural language processing (NLP), and predictive analytics. These areas are vital for various healthcare applications.

11.2.4.1 Convolutional Neural Networks

Convolutional Neural Networks (CNNs) are highly specialized deep learning models that excel in processing structured grid data, particularly images. Convolution is a mathematical approach used to efficiently identify patterns and characteristics in photographs, considering the spatial hierarchy of these features.

Normally, CNNs work with medical images such as X-rays, CT scans, and MRIs to detect diseases, anomalies or temporal variations. They can recognize patterns not visible to human eyes hence making it easy for a doctor to diagnose fast and plan a treatment quickly. Skin cancer classification can be improved by CNN inspecting images showing skin abnormalities. According to experts, AI and ML should be embraced by health personnel in their fields while considering various aspects such as the constraints among others when making diagnoses plans for treatments or even caring patients however these machines process information like human beings only through knowledge of medicine and use analytic data as well as such cognitive theories based upon health principles.

11.2.4.2 Long Short-Term Memory

Long short-term memory (LSTM) recurrent neural networks (RNN) are a subclass of RNNs that were specifically designed to address the problem that conventional RNNs fall short when handling long-term sequences. However, LSTMs typically suffer from disappearing gradients issues, which cause them to forget what was learned in the beginning of the sequence as it becomes too long. This problem was tackled by LSTMs through the development of a new architecture that makes it possible for them to keep track of information over many time steps hence applicable in healthcare when working with sequences such as medical records, signals processing or genomic.

This includes both an input, output and forget gates along with a cell state that controls how much data is kept within the cell and what should be forgotten or retained thereby allowing erasure or storage of memories when necessary.

Input Gate: It determines how much new information coming into the cell state should be concocted or allowed in.

Forget Gate: It decides what information remains in and what needs to be let go from the cell state.

Output Gate: It helps to release specific part of the cell state into output dataset.

Due to their ability to effectively capture far-field interactions between different subsets of an input sequence depending on context real-time patient monitoring would require EHRs coupled with wearable devices along with lab tests about systems build based on LSTMs specifically for complex sequential processing tasks usually encountered within health settings.

LSTM applications such as analyzing patterns of heart failure or sepsis in time series data can lead to early interventions. Additionally, LSTM can be utilized for pattern recognition or anomaly detection in physiological signals including ECGs and EEGs. Specifically, tiller chains in EEG traces can be singled out by LSTM models while ECG signals help identify abnormal heart rhythms thus making early diagnosis possibility with treatment planning handy. When it applies to medical natural language processing tasks, LSTM stands out head and shoulders above everything else; these tasks encompass extracting patient information from clinical notes, evaluating patient responses, and even automating doctor-patient conversation transcriptions. They grasp the significance and subtlety of tone in text, resulting in improved accuracy in sentiment analysis and text extraction. Lastly, for drug discovery purposes, LSTMs can utilize chemical and biological data sequences to predict substances' activity levels or simulate new drug production processes. Genomics refers to the study of cells' genes which are used to identify genetic changes responsible for diseases hence enabling targeted drug discovery process.

The incorporation of AI elements within clinical workflows may significantly enhance healthcare provision (for instance, offering real-time analytics, improving

TABLE 11.4
AI Capabilities Designed to Assist in Clinical Decision-Making

AI Functionality	Application in Clinical Settings	Benefits to Clinical Decision-Making
Predictive Analytics	Predicting patient deterioration by analysing vital signs in real-time	Early intervention and tailored care
Natural Language Processing (NLP)	Extracting patient information from unstructured clinical notes	Improved data accessibility and patient history comprehension
Image Recognition and Analysis	Diagnosing diseases from medical images (e.g., X-rays, MRI)	Enhanced diagnostic accuracy and speed
Personalized Treatment Recommendations	Analysing genetic data to recommend personalized medication	Optimized treatment plans tailored to individual genetic profiles
Robot-assisted Surgery	Assisting surgeons with precise movements during operations	Increased surgical precision and reduced recovery times
Automated Alerts and Reminders	Monitoring patient medication adherence and sending reminders	Improved patient compliance and monitoring
Chatbots for Patient Triage	Initial patient assessment and triage via online platforms	Efficient patient routing and reduced wait times
Electronic Health Record (EHR) Analysis	Identifying patterns in patient data for risk assessment	Enhanced identification of at-risk patients for proactive care

diagnostic accuracy rates as well as treatment specificity, and tailoring individuals' functionality) [see section 11.4]. Additionally, it offers some sophisticated decision-making instruments to this group of professionals that work at hospitals and other medical facilities aiming at improving patients' treatment options and operational effectiveness.

11.3 APPLICATIONS OF AI IN HEALTHCARE

The numerous transformations in the medical sector, including patient care improvements and the new approaches used in diagnosing diseases and their treatments are among the outcomes of AI [9]. This section gives examples of important spheres where AI is used in healthcare with supporting statistics on its effects up to the near future if more resources are invested into it.

11.3.1 DIAGNOSTIC IMAGING

1. **Statistical Data:** The involvement of AI in radiology-based tasks for diagnosis over time has resulted in an increase of up to 30% accuracy percentage together with pathology tests using this paradigm.
2. **Analysis:** Diagnostic imaging systems power by artificial intelligence can analyze images more quickly than earlier methods but they also achieve higher precision thanks to deep learning algorithms which can pinpoint anomalies such as malignant cells on CT scans then segregate them (for example detecting cancer) or diagnose diabetic retinopathy from retinal pictures [10]. These improvements hasten the time of diagnosing while boosting accuracy hence better patient outcomes.

11.3.2 PREDICTIVE ANALYTICS FOR PATIENT MONITORING

1. **Statistical Data:** Within a specified duration lower rate in terms of percent was recorded in hospitals during readmissions after using AI for patient monitoring.
2. **Analysis:** Using real-time wearable device data as well as electronic health records (EHR), artificial intelligence models are capable of foreseeing adverse outcomes than traditional methods because they operate faster [11]. In cases with chronic diseases like heart illness this could be very helpful since before it's too late it can detect all signs of deterioration hence patients' life quality improves while they're spared rehospitalization.

11.3.3 PERSONALIZED MEDICINE

1. **Statistical Data:** In terms of receiving treatment based on genetic makeup individual variation rates, the number has gone up by 40% because AI helps personalize these procedures.
2. **Analysis:** This enables the development of individualized treatment plans taking into account patients' specific needs through analysing genetic

information alongside medical history files plus current health statistics using artificial intelligence algorithms [12]. It does save lives which is particularly applicable to the field oncology—where treatments can be much more effective for certain individuals because of the genetic composition that influences their response to the same drugs used by others.

11.3.4 DRUG DISCOVERY AND DEVELOPMENT

1. **Statistical Data:** AI has cut the average time taken for drug discovery and development down from 12 years to 6 years.
2. **Analysis:** The machine-learning based on big data analysis techniques can identify potential drugs much faster than conventional methods [13]. Moreover, these systems predict their efficacies and safety profiles accurately better than the same characteristics made by traditional mediums at the moment of their inception so new drugs appear in the market in a shorter period thus reducing costs of their release making them less expensive.

11.3.5 VIRTUAL HEALTH ASSISTANTS

1. **Statistical Data:** The use of remote medical services by stand-in doctors is associated with a 30% reduction in unnecessary visits.
2. **Analysis:** Due to chatbots and virtual assistants that are AI-powered, they can respond promptly 24*7, by giving medical consultations based on the clients' symptoms, remind about the necessity; to take pills or other treatments, provide emotional support when it is deemed necessary [12]. This encourages patient engagement thereby freeing healthcare providers to attend to more urgent needs.

11.3.6 ROBOTIC SURGERY

1. **Statistical Data:** Utilizing AI in robotic surgery has increased accuracy levels during operations leading to a 70% improvement in precision while reducing recovery periods by up to 40%.
2. **Analysis:** On the other hand, AI is capable of enabling surgical robots perform complex procedures with highest level of accuracy and minimum invasiveness [14]. Apart from improving surgical outcomes, this approach also cuts down on patient recovery time significantly.

Figure 11.2 compares the use of AI in healthcare across different parts of the world. It is led by North America, followed by Europe and Asia-Pacific. India has a 25% adoption rate, which puts it ahead of Latin America and the Middle East & Africa but behind other high-performing regions [15]. This increase in interest and investment from India towards AI technology for healthcare can be attributed to its booming tech industry as well as government initiatives towards digital health advancement. The disparity between how much different regions integrate AI into

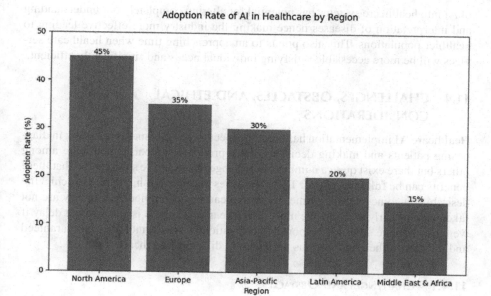

FIGURE 11.2 Adoption rate of AI in Healthcare region.

their healthcare systems reflects economic, technological and policy factors that come into play with this kind of undertaking. For instance, higher adoption rates seen in North America or Europe may be because they have strong digital infrastructure, heavy investments on AI research & development and favourable regulatory environments [16].

With these "average adoption rates," significant participation occurs, except in leading countries like India. This is due to the presence of a strong IT sector, alongside initiatives like the National Digital Health Mission (NDHM), which aims at transforming healthcare to digital forms across India. The take up of AI in Latin America and Middle East Africa (MEA) regions is lower than other global regions. This can be attributed to low financial resources among populations inhabiting these areas due to limited legislative policies hindered by less infrastructure development projects among others. Even though there are challenges in these areas, governments and commercial organizations are more aware of how valuable AI is particularly within healthcare; they have begun setting up extensive disease surveillance networks and making health services easily accessible to people living far away from cities [16].

The histogram in Figure 11.2 summarizes how much artificial intelligence is being used in global medical care today, and its current level of implementation across various regional contexts. These statistics show how much we can achieve by using this technique for diagnosis precision improvement process, personalizing treatment plans according to certain patient characteristics thereby fast-tracking drug finding process and improving patient care as well as outcomes. Additionally, introduction of AI into healthcare systems has potential for altering completely our understanding and interpretation of diseases hence making the industry more effective leading to healthier populations. This also points to an approaching time when healthcare services will be more accessible satisfying individual needs and significantly efficient.

11.4 CHALLENGES, OBSTACLES, AND ETHICAL CONSIDERATIONS

Healthcare AI implementation has great prospects when it comes to assisting in diagnosing patients and making decisions about proper treatment, patient care among others but there exist quite a number of challenges that require solutions so that these benefits can be fully harnessed. These obstacles may act as a hindrance to achieving desirable outcomes within a time frame that can be accommodated, if they are not taken care of early enough. We must therefore address these issues without delay if we wish to make sure that responsible utilization of AI technologies is guaranteed and fairness in their applications promoted in different healthcare situations.

11.4.1 CHALLENGES AND OBSTACLES

1. **Data Privacy and Security:** For AI systems to operate effectively on a large scale they require extensive personal data, such as private medical histories. However, preserving the confidentiality and security of this data is vital because unauthorized access or invasion of privacy may be harmful. However, adherence to regulations like GDPR (General Data Protection Regulation) within Europe; or HIPAA (Health Insurance Portability and Accountability Act) which deals with patient information within the United States could pose some problems due to complexity and detail associated with them. According to the Ponemon Institute survey findings, 89% of the companies had experienced at least one case of data breach over the last two years while 41% had experienced cases that exceeded five during this period [17]. At the moment data safety is one of the major challenges in the healthcare industry.

2. **Bias Data and Inequality:** The results produced by AI models are dependent on the quality of the training data; if biases exist or diversity is lacking in these datasets then unfair treatment outcomes could arise for patient groups with different attributes. These health disparities may deepen existing social determinants of health problems among marginalized populations. Science journal published a research paper showing that an AI system used in healthcare had lower tendency of recommending higher care levels for African Americans than Whites [18]. This happened because the program wrongly assumed that blacks were healthier than they were, therefore indicating how biased training data affects patient care.

3. **Integration within Clinical Workflows:** There are logistical and technical challenges involved in integrating artificial intelligence capacities into current healthcare systems and processes. Healthcare providers need skills on how best these tools can be used while they also have to be compatible with existing healthcare information technology infrastructure [19]. A poll

carried out by AMA last year found that over 85% of doctors recognize prospects for improving patient care through digital health solutions but only about 28% integrate such technologies into their practice implying significant barriers exist [20].

4. **Algorithm Transparency and Explainability:** Some AI models especially those based on deep learning are often called "black boxes" due to their complex nature which makes it hard understand why certain decisions were made or even question them thus creating trust issues between caregivers/clinicians and patients. A study released in Nature Machine Intelligence revealed that more than 70% of healthcare workers feel the need for greater transparency in AI systems to trust them and use them effectively for clinical decision-making [21].

5. **Regulatory Challenges:** The rapid advancement of AI poses challenges for regulators to set and enforce appropriate standards that ensure the safety, efficacy, and ethical use of AI in healthcare without stifling innovation. By 2021, the US Food and Drug Administration (FDA) had approved over 130 AI-based medical devices and algorithms [22]. This reflects the growing but complex landscape of AI in healthcare that requires careful regulatory oversight.

11.4.2 ETHICAL CONSIDERATIONS

1. **Patient's Consent:** Patients must be informed of their entitlements concerning the utilization of AI systems in healthcare and mandated to provide explicit consent for its utilization. The ethical quandary of upholding patient autonomy in an era where computers possess the capability to make or heavily influence decisions is intricate. Based on a survey conducted by Ipsos on behalf of the World Economic Forum, 73% of individuals globally express concern about the potential misuse of their personal data by AI systems [23]. This indicates that it is necessary to gain informed permission while also preserving individual liberty in the context of medical care driven by artificial intelligence.

2. **Accountability:** Assigning accountability for the consequences of AI-driven assessments on patients is challenging due to the complexity of the underlying issues. The issue of accountability for decisions made by AI systems in hospitals, whether it lies with developers, healthcare providers, or the AI systems themselves, needs to be clarified. An Accenture survey found that 81% of health executives concur with the notion that "AI will fundamentally revolutionize the way they obtain information from and interact with patients" [24], hence raising concerns about the responsibility for judgments made by artificial intelligence.

3. **Equity and Access:** The use of artificial intelligence in health has potential global impact on health outcomes; however, it could also amplify the gap between those who can access advanced medical technology and those who cannot do so currently Ensuring fairness regarding access rights vis-a-vis benefits emanating from artificial intelligence in healthcare is an ethical imperative. The World Health Organization's first global survey on AI in

health, conducted in 2021, points to the potential of AI worsening health disparities if not well regulated [25]. In particular, there are fears that developing countries may be left behind during the AI revolution.

4. **Patient-Doctor Relationship:** There is a risk that increasing reliance on AI could alter traditional doctor-patient dynamics. While AI has the potential to enhance diagnostic and therapeutic capabilities, it is important for humanity's side of healthcare to be preserved through empathy, understanding, and holistic patient assessment. A *New England Journal of Medicine* Catalyst Insights Council study found that 58% of clinical executives saw diagnostic formulation as the area where AI had greatest promise [26]. However, concerns have been raised about dehumanization by machines taking over some aspects of provider-patient interaction in healthcare settings.

A worldwide system for healthcare is required to cater for these problems and solve ethical issues through cooperation among scientists, healthcare providers, ethicists patients as well as policymakers. Ensuring data security, inclusive training data sets, transparency in AI decision making and explicit regulation and ethical framework are necessary steps towards responsible integration of AI in healthcare. Ongoing debates, research, and policy developments are essential in dealing effectively with the challenges posed by artificial intelligence systems while maximizing their potentiality of enhancing patient care experiences and health outcomes.

11.5   FUTURE TRENDS IN AI AND HEALTHCARE

The future of healthcare AI has the potential to upend everything about how we treat patients, from diagnosis to treatment and management; it may also bring a lot of good into medicine. As machines get smarter so will their use within our hospitals, thus making them more efficient at what they do best, which is treating people who need help getting better. Here are the future possibilities for healthcare AI:

11.5.1   PERSONALIZED MEDICINE AND GENOMICS

By analyzing massive datasets, AI can speed up the process of personalized medicine by better matching therapies with specific genetic traits. More precise targeting medicines against diseases like cancer may be possible with the help of AI and genomics [27]. This might include several things, such as the ability to detect genetic changes and predict how a patient will react to a drug.

11.5.2   PREDICTIVE ANALYTICS FOR PREVENTIVE CARE

Implementing predictive analytics using AI allows medical personnel to predict different diseases' outbreak before they occur; recognize patient's decline rate earlier in order to act promptly in correcting it, anticipate upcoming health disasters in advance such as heart attacks or stroke among others. Transitioning from the current reactive treatment models adopted in many parts of the world into proactive prevention by means of prediction would significantly enhance public health results.

11.5.3 ADVANCED DIAGNOSTIC TOOLS

As good as diagnostic tools may currently be, artificial intelligence will improve their precision and increase their rate. More simply put, using deep learning algorithms properly can assist physicians by giving them fast results when studying medical images, pathology slides, etc., for cancer tumors, cardiovascular diseases or even brain disorders such as Alzheimer's.

11.5.4 ROBOTIC SURGERY AND ASSISTANCE

Robotic surgeons enabled by AI can also carry out more precise surgeries, which in turn will help patients recover faster from their procedures. Further, in conjunction with human therapists, robots can perform physical therapy, which would enhance the rehabilitation process for patients and speed up their recovery [28, 29].

11.5.5 VIRTUAL HEALTH ASSISTANTS AND TELEMEDICINE

Among the various functions of AI-powered virtual health assistants is the provision of individualized recommendations regarding dietary habits, exercise routines, and other lifestyle factors that have an impact on a person's health. These programs can provide mental healthcare when needed while also continuously monitoring chronic conditions. Doctors can treat patients in faraway places without having to be in the same room through AI-driven telemedicine. This has led to an increased provision of quality healthcare to individuals residing in remote areas.

11.5.6 HEALTHCARE SYSTEM EFFICIENCY

AI has multiple methods of expediting efficiency in the healthcare system such as patient intake scheduling, supply chain management and invoicing among other numerous ways. In doing so, it saves time spent by administrators manually handling these duties enabling them to concentrate further on providing care to the patients directly. At the same time, the overall interaction between patients and providers will be enhanced largely due to ease that come with dealing with electronic systems as opposed to today's processes that are based on paperwork where mistakes are common leading to misunderstandings between parties on either side hence undermining quality of service provided.

11.5.7 ETHICAL AI DEVELOPMENT AND USE

Given the growing awareness of ethical concerns about AI use in healthcare, ethicality in its application must improve. AI-powered medical service delivery must always be fair for example. Besides, patients' privacy rights must invariably remain protected to developers as well as policymakers. Therefore, to make it clear for doctors or nurses or other staffs about how different systems made specific decisions and at the same time make these mechanisms available to patients themselves there should be an assurance that there is transparency, accessibility and understand ability in algorithms used.

11.6 CONCLUSION AND FUTURE SCOPE

Most of the work in this area has studied artificial intelligence and machine learning as well as their basic principles, healthcare applications; possible drawbacks; ethical considerations and remedies. The session further looked at the basic concepts behind AI, ML, types of AI and ML, as well as the terminologies used to describe them. This information is very important if one wants to understand the applicability of these technologies in medicine. Machine learning technologies and artificial intelligence have the potential to greatly improve the treatment of patients. However, there a number of challenges that need to be overcome before these technologies can be widely used in clinical settings. Such tasks include securing patient data, removing algorithmic bias, embedding them within clinical workflow and ensuring any consent by any patient group is fair to them all. They are the main agents of digital transformation in the field of health. This calls for them embracing technology in their day-to-day activities. Therefore, it is imperative that technologists, ethicists, politicians, and other players involved in service delivery should work collaboratively. Regular training on such tools is also necessary because they keep on changing.

This research has led us to many questions, such as how patient privacy could be better ensured in the AI age. Discovering methods to identify and fix bias in AI systems by diversifying and inclusively using datasets is another intriguing topic. How can healthcare providers incorporate AI-based procedures and technologies into their existing infrastructure without significantly impacting patient care or laying off employees? Lastly, what kind of ethical frameworks should be put in place to ensure fairness when it comes to these machines? We may find ourselves in an unfair situation where certain individuals have access solely through machines while others, who may require it just as much or even more, do not. This is because humans are better at communicating when necessary, while machines have limitations like not being able to understand emotions as well. Going forward, a number of future study questions are evident. How can we ensure the confidentiality of patient information in the age of machine learning? How can we identify and fix AI systems that have algorithmic biases? Without negatively impacting patient outcomes or interfering with medical practices, which models would best serve to integrate artificial intelligence into current healthcare services? In regulating the creation, deployment, and utilization of AI in healthcare, what ethical frameworks can guarantee equity while preventing the worsening of current health disparities? We need further studies to determine how ML and AL affect healthcare costs, operational efficiencies, and patient outcomes.

REFERENCES

[1] M. Gupta, S. Ahmed, R. Kumar, and C. Altrjman (Eds.). Computational Intelligence in Healthcare: Applications, Challenges, and Management. CRC Press, 2023.

[2] R. S. Lee and R. S. Lee, "AI fundamentals," Artificial Intelligence in Daily Life, pp. 19–37, 2020.

[3] A. C.-C. Liu, O. M. K. Law, and I. Law, Understanding Artificial Intelligence: Fundamentals and Applications. John Wiley & Sons, 2022.

[4] J. Raffort, C. Adam, M. Carrier, and F. Lareyre, "Fundamentals in artificial intelligence for vascular surgeons," Annals of Vascular Surgery, vol. 65, pp. 254–260, 2020.

[5] K. Yao and Y. Zheng, "Fundamentals of machine learning," in Nanophotonics and Machine Learning: Concepts, Fundamentals, and Applications, Springer, pp. 77–112, 2023.

[6] J. M. Cherian and R. Kumar, "Fundamentals of machine learning," in A Guide to Applied Machine Learning for Biologists, Springer, pp. 147–174, 2023.

[7] A. Oza and A. Bokhare, "Diabetes prediction using logistic regression and K-nearest neighbor," in Congress on Intelligent Systems: Proceedings of CIS 2021, Volume 2, Springer, pp. 407–418, 2022.

[8] S. Razavi, "Deep learning, explained: Fundamentals, explainability, and bridgeability to process-based modelling," Environmental Modelling & Software, vol. 144, p. 105159, 2021.

[9] J. D. Rodríguez-García, J. Moreno-León, M. Román-González, and G. Robles, "Introducing artificial intelligence fundamentals with LearningML: Artificial intelligence made easy," Eighth International Conference on Technological Ecosystems for Enhancing Multiculturality, Diagnostic Pathology, pp. 18–20, 2020.

[10] S. Shafi and A. V. Parwani, "Artificial intelligence in diagnostic pathology," Diagnostic Pathology, vol. 18, no. 1, p. 109, 2023.

[11] A. İ. Tekkeşin, "Artificial intelligence in healthcare: Past, present and future," Anatolian Journal of Cardiology, vol. 22, no. Suppl 2, pp. 8–9, 2019.

[12] R. G. Curtis et al., "Improving user experience of virtual health assistants: Scoping review," Journal of Medical Internet Research, vol. 23, no. 12, p. e31737, 2021.

[13] D. Paul, G. Sanap, S. Shenoy, D. Kalyane, K. Kalia, and R. K. Tekade, "Artificial intelligence in drug discovery and development," Drug Discovery Today, vol. 26, no. 1, p. 80, 2021.

[14] M. Bhandari, T. Zeffiro, and M. Reddiboina, "Artificial intelligence and robotic surgery: Current perspective and future directions," Current Opinion in Urology, vol. 30, no. 1, pp. 48–54, 2020.

[15] U. J. Muehlematter, P. Daniore, and K. N. Vokinger, "Approval of artificial intelligence and machine learning-based medical devices in the USA and Europe (2015–20): A comparative analysis," The Lancet Digital Health, vol. 3, no. 3, pp. e195–e203, 2021.

[16] B. X. Tran et al., "Global evolution of research in artificial intelligence in health and medicine: A bibliometric study," Journal of Clinical Medicine, vol. 8, no. 3, p. 360, 2019.

[17] J. M. Puaschunder, "The future of artificial intelligence in international healthcare: An index," Proceedings of the 17th International RAIS Conference on Social Sciences and Humanities, Scientia Moralitas Research Institute, pp. 19–36, 2020.

[18] N. Norori, Q. Hu, F. M. Aellen, F. D. Faraci, and A. Tzovara, "Addressing bias in big data and AI for health care: A call for open science," Patterns, vol. 2, no. 10, 2021.

[19] K. Zhai, M. S. Yousef, S. Mohammed, N. I. Al-Dewik, and M. W. Qoronfleh, "Optimizing clinical workflow using precision medicine and advanced data analytics," Processes, vol. 11, no. 3, p. 939, 2023.

[20] D. J. Blezek, L. Olson-Williams, A. Missert, and P. Korfiatis, "AI integration in the clinical workflow," Journal of Digital Imaging, vol. 34, pp. 1435–1446, 2021.

[21] A. Moltubakk Kempton and P. Vassilakopoulou, "Accountability, transparency and explainability in AI for healthcare," 8th International Conference on Infrastructures in Healthcare, Radiation, 2021.

[22] F. Pesapane et al., "Legal and regulatory framework for AI solutions in healthcare in Eu, US, China, and Russia: New scenarios after a pandemic," Radiation, vol. 1, no. 4, pp. 261–276, 2021.

[23] B. Pickering, "Trust, but verify: informed consent, AI technologies, and public health emergencies," Future Internet, vol. 13, no. 5, p. 132, 2021.

[24] I. Habli, T. Lawton, and Z. Porter, "Artificial intelligence in health care: Accountability and safety," Bulletin of the World Health Organization, vol. 98, no. 4, p. 251, 2020.

[25] C. R. Clark et al., "Health care equity in the use of advanced analytics and artificial intelligence technologies in primary care," Journal of General Internal Medicine, vol. 36, pp. 3188–3193, 2021.

[26] S. Segers and H. Mertes, "The curious case of 'trust' in the light of changing doctor—patient relationships," Bioethics, vol. 36, no. 8, pp. 849–857, 2022.

[27] K. B. Johnson et al., "Precision medicine, AI, and the future of personalized health care," Clinical and Translational Science, vol. 14, no. 1, pp. 86–93, 2021.

[28] S. Bodenstedt, M. Wagner, B. P. Müller-Stich, J. Weitz, and S. Speidel, "Artificial intelligence-assisted surgery: Potential and challenges," Visceral Medicine, vol. 36, no. 6, pp. 450–455, 2020.

[29] G. M. Minopoulos, V. A. Memos, K. D. Stergiou, C. L. Stergiou, and K. E. Psannis, "A medical image visualization technique assisted with ai-based haptic feedback for robotic surgery and healthcare," Applied Sciences, vol. 13, no. 6, p. 3592, 2023.

12 Early Ethical Considerations and Societal Impacts on Gender Health

*Antonitta Eileen Pious, K. Ananthi,
Mary Fabiola, and Benedict Tephila*

12.1 INTRODUCTION

Data in medicine mostly focuses on the male section of the society. Only after 25 years was a law put forth to ban this criteria and include women in biomedical research. Male bias pervades the medical industry from the data collected, tested to the interventions made and benefits provided. Artificial intelligence helps remove this disparity by looking into details from the large datasets provided and helping make apt decisions as to what changes can be made or taken to avoid such gender related issues.

One key feature of this ethical framework is that it examines the details involved in early societal impact on gender health. It reflects on how gender coincides with other social determining factors like the individual's experiences, vulnerabilities, and healthcare needs. For example, transgender individuals across Asia may face barriers to having access to gender-affirming care due to the country's system's discrimination, their own economic background status, or lack of culturally competent providers in the healthcare field. By being aware of these intersections, healthcare professionals can better provide a reliable and honest means of healthcare to these under-biased parts of the society.

Moreover, the ethical protocol highlights the importance of addressing structural inequalities and cultural biases inside the healthcare systems. This unethical means of being biased has only led to the downfall of the society as a whole, as the citizens start to believe that their race, gender or social status prevent them from having equal access to rights. Cultural biases and stereotypes can also shape healthcare providers' attitudes and behaviors towards patients, leading to terrible differences in treatment and quality of care. By interrogating and challenging these systemic barriers, the ethical protocol aims to promote fairness, justice, and diversity and inclusion within healthcare practice.

Ethical protocol may include:

1. **Culturally competent care:** Requesting providers to undergo training on how to handle cultural humility and competency to gain a better understanding and address the unique needs and perspectives of heterogeneous patient populations.

2. **Intersectional analysis:** Encouraging healthcare practitioners to consider how to join social identities, such as gender, race, and socio status, influence patients' experiences and healthcare outcomes.
3. **Fair resource allocation:** Fair resource allocation to patients irrespective of where and how they come from or are even belonging to the backward section of the society.
4. **Patient-focussed care:** Involvement of patients in decision making and respecting their autonomy, preferences, and values in healthcare delivery.
5. **Advocacy and activism:** Healthcare workers are being asked to identify and advocate for changes to the system that will fix structural inequalities and improve everyone's health, no matter their gender or other social identity.

Figure 12.1 show cases the development of an ethical protocol which encourages healthcare advocates to foster and calibrate a system that addresses for systemic changes that takes into account the inequalities and promote health equity for all human beings, regardless of their gender or other social identities. A protocol to lead the analysis of gender-related ethical problems in healthcare settings represents a crucial step towards promoting equity, justice, and inclusivity within healthcare practice. By understanding the intersectionality of gender with other social determinants of health and addressing structural inequalities and cultural biases, this protocol produces a more ethical and compassionate way to relate to individuals of all diverse identities and backgrounds.

By including ethical principles with sociological insight, our framework throws light into a complete approach of addressing gender differences. It acts as an important tool for AI methodologies to support healthcare practitioners, policymakers, and researchers to thoroughly examine and rectify early ethical considerations and societal impacts on gender health. By advocating gender-inclusive analysis and advocacy, it contributes to the upliftment of the women in our society.

1. **Transgender-access to healthcare:** In many healthcare systems, transgender individuals face many barriers to accessing gender-affirming care, including treatments like hormone therapy and gender confirmation surgery. A real-time example involved a transgender woman who was denied coverage for gender confirmation surgery by her insurance provider, stating it was a cosmetic procedure. This case reflects the system where a required treatment was not provided because of a socially incompetent and ignorant provider. A recent survey in Netherlands depicts 67% of adult Trans had to wait more than 18 months for a basic initial gender check up and more than 6 months for a surgical check up.
2. **Maternal mortality disputes:** In the United States, Black women are very much affected by childbirth mortality compared to white women. A real case involved a Black woman who died shortly after giving birth due to complications that were not sufficiently taken care of by healthcare providers. This case shows the influence of race and gender in healthcare outcomes and highlights the need to address systemic biases and disparities in maternal healthcare.
3. **Marginalized women's access to reproductive health services:** In many regions, women from backward communities, such as low-income women

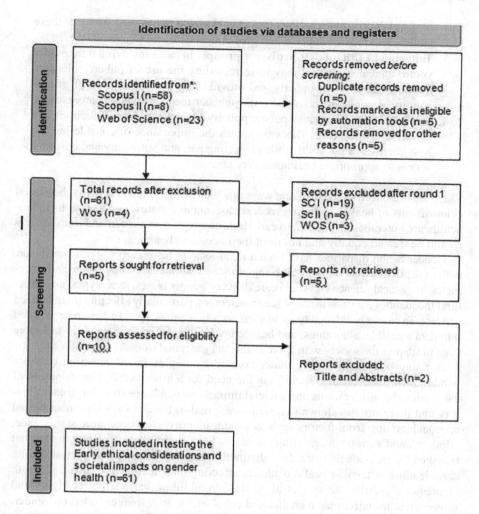

FIGURE 12.1 Tracking gender ethics: tracing the path from literature search to ethical development.

and women of color, face barriers to accessing sexual and reproductive health services, including contraception and abortion care. An example involved a young woman of color who encountered extreme judgment from healthcare providers when seeking contraception services, leading to delays in accessing care. Concerns of patient personal freedom, confidentiality, and dignity in reproductive healthcare are highlighted by this case.

4. **Care for victims of trauma and gender-based violence:** Violence occurs in about 35% of women globally. In India of a count of 10,000 women 26% are assaulted physically by their own spouses. Intimate violence from partners result in mental trauma, sexually transmitted diseases and various other diseases both mentally and physically. It is considered the duty

of healthcare systems to take into consideration the mental status of these victims and provide support and give utmost care and treatment.

5. **Tailoring health models to diverse groups:** In cases involving transgender youth, ethical dilemmas may arise regarding the use of puberty blockers to delay the onset of puberty and provide time for exploration of gender identity. A real case involved a transgender teenager whose parents were in a confusion whether to pursue puberty blockers as part of their child's gender-affirming care. This case shows the importance of considering the best interests of the child while respecting parental autonomy and ensuring access to appropriate healthcare services.

These real cases tell us the different ways in which gender intersects with other social determinants of health, including race, socioeconomic status, and age, to influence healthcare outcomes and access to care. It throws light that women and transgenders should be treated equally and not limit their access to healthcare.

Gender health disparities have been a main issue in healthcare systems worldwide, with in depth thoughts for the well-being of individuals and societies. Despite advancements in medical science and healthcare delivery, gender biases, stereotypes, and structural inequalities continue to shape health outcomes, particularly affecting marginalized genders. To ensure health equity for women, society requires a gender lens to be included into data available, algorithms, and healthcare. Healthcare professionals can feel more keen to support the society with apt relevant data generated from AI algorithms.

A rising body of literature underscores the complex interconnection between gender, health, and ethics, specifying the need for a more detailed examination of early ethical considerations and societal impacts on gender health. For decades history and literature has shown that Healthcare realms have always shown an biased approach arising from factors such as gender identity, socioeconomic status, race, ethnicity, and various intersecting variables. For example, research has shown that transgender individuals often face distinct barriers to accessing gender-affirming care, leading to terrible health disasters affecting their mental and physical health (James et al., 2016; Reisner et al., 2016). In addition, various policy rights and government initiatives have challenged equal access to resources based on gender, and this remains an ongoing debate today (Shaw, 2018; Zampas & Gher, 2018).

Moreover, while respecting rights and cultural values, early proposals in gender health included important ethical considerations, such as complex moral questions about autonomy, justice, self-independence, prenatal screening, and gender-affirming treatments, as well as the role of healthcare professionals in these decisions (Baldwin et al., 2018; McDougall, 2020a).

The institutional policies and practices in America perpetuate a discrimination against minority groups because of unjustified opinions similar to microaggression related to racism (Alang et al., 2020; Bowleg, 2012).

In order to overcome these obstacles, scholars and eminent professionals have advocated for a more diverse and inclusive strategy to address gender health differences, one that acknowledges the interdependence of social, cultural, and economic variables influencing health outcomes. Researchers can create more dynamic frameworks for acknowledging societal impacts on gender health and early ethical considerations by integrating insights from gender studies, sociology, and bioethics. This will help the overall development of the society and not let the healthcare industry govern in an unbiased manner.

12.2 OBJECTIVE

The main aim of this study is to develop a comprehensive theoretical framework that can help examine the first moral dilemmas and societal repercussions associated with gender health in the healthcare industry.

The objective is to:

1. Check on the complex moral conundrums that arise from early interventions and gender-related healthcare practices, such as prenatal care, reproductive health services, and gender-affirming medical procedures.
2. Examine the ways that institutional regulations, cultural norms, and access to healthcare resources are only a few examples of the societal and structural injustices that affect the health of women.
3. Develop an ethical protocol adapted to analyze and address gender-specific ethical issues in health services, incorporating aspects of autonomy, justice and equality.
4. Test and refine the developed ethical protocol through case studies or an empirical review of the relevant literature, with a particular focus on early interventions and their impact on sexual health.
5. Provide knowledge and recommendations from AI methodologies to health professionals, policymakers, and researchers to improve gender-specific practices and reduce disparities in healthcare, ultimately promoting health equity and social justice among all genders.

The research aims to increase understanding of the ethical complexities of achieving these goals and the social impact of sexual health, providing actionable strategies to promote a gender-responsive health environment and promote the well-being of diverse populations.

12.3 LITERATURE SURVEY

Gender health disparities have long been considered a critical public health problem. Research has documented significant differences in health outcomes, access to healthcare, and quality of care based on gender identity and expression (Johnson et al., 2014; Bowleg, 2012). A large body of evidence suggests discrimination is a critical psychosocial stressor that can navigate the path to negative changes in health status and altered behavioral patterns. (Williams & Mohammed, 2013). Transgender people may experience barriers to seeking specific gender-specific healthcare because of high prejudice, peer disapproval, and not sufficient well-informed health professionals (Grant et al., 2011; Reisner et al., 2016). Women from marginalized communities often experience significant mental trauma due to feelings of neglect, which contributes to higher rates of maternal mortality and increased vulnerability to certain diseases (Krieger, 2012; Howell et al., 2016). Principles of oneness, justice, beneficence and non-criminality guide ethical decisions in healthcare. Ethical involvement plays a crucial role in eliminating gender health disparities and promoting equal healthcare (Beauchamp & Childress, 2013). But applying these ethical rules require the respecting of one's private details and fidelity (Feder & Kimport, 2012; McDougall, 2020). In addition to this situation, ethical challenges emerge regarding the provision of gender-specific treatments, including issues with ensuring informed consent, balancing medical interventions with patient autonomy, and addressing disparities in access to necessary care for different genders (Bauer et al., 2009; Baldwin et al., 2018). To prevent the potential stigmatization of marginalized communities or gender-based discrimination during physical health examinations provided by health authorities, fostering respectful and culturally sensitive interpersonal communication is essential. This includes healthcare providers actively engaging in dialogue with patients, showing empathy, and building trust through mutual understanding. By creating a safe, non-judgmental environment where patients feel heard and respected, healthcare professionals can reduce the risk of stigmatization, encourage open discussion of health concerns, and promote equitable treatment across all communities, regardless of gender or background. This approach helps ensure that care is delivered without bias, leading to better health outcomes and patient satisfaction (Goffman, 1963; Kuhlmann et al., 2017). The lack of insurance coverage for gender-specific procedures, coupled with restrictive legislation, creates significant structural barriers that limit transgender and nonbinary individuals' access to essential healthcare services. Many insurance plans fail to cover critical medical interventions such as hormone therapy, gender-affirming surgeries, or mental health support specifically tailored to the needs of these populations. Additionally, restrictive laws in some regions may hinder access to gender-affirming care or impose burdensome requirements, such as psychological evaluations or waiting periods, before treatment can begin. These structural barriers not only delay necessary medical interventions but also contribute to health disparities, increasing the vulnerability of transgender and nonbinary individuals to mental health issues, discrimination, and poor overall health outcomes. (Lombardi et al., 2001; James et al., 2016). Health systems may integrate gender bias and inequality through not sufficient training, insensitive practice, and heteronormative assumptions (Hoffman et al., 2014; Whitehead et al., 2016). Addressing this complex issue requires the development of a gender-sensitive framework within healthcare systems. The famous intersectionality theory, proposed by Crenshaw (1989), throws light on the interconnection of social identities, including gender, race, class, and sexuality, in developing individuals' experiences and health outcomes. Creating an intersectional lens and gender lens to health ethics enables a more intricate understanding of how multiple axes of oppression intersect and are linked to health disparities (Bowleg, 2012; Hankivsky, 2012). Furthermore, feminist ethics provides a critical lens on traditional ethical theories, highlighting the importance of relativity, the ethics of care, and the inclusion of marginalized voices in ethical decision-making (Gilligan, 1982; Tong, 1993). In the context of sexual health, several recommendations have been put forward to address these concerns effectively. These include including gender competency education in health curricula, promoting inclusive healthcare environments that affirm diverse gender identities, and promoting policy changes to ensure equal access to gender-affirming care (American Medical Association, 2018; World Health Organization, 2019). In addition, promoting patient-centered therapies that prioritize patient autonomy, informed decision-making, and respectful communication can improve healthcare for all genders (Institute of Medicine, 2001; Joint Commission, 2001).

12.4 RESULTS

Sixty-one abstracts were obtained from the literature review published in the years 1996–2023. Abstracts were systematically analyzed using Braun and Clarke's six-step thematic analysis framework. This process involved reading and coding abstracts, developing initial themes, discussing those themes, reaching agreement on identified themes, and finally naming final themes. Inspired by Braun and Clarke's methodology, observations were made visually using a mind map (see Figure 12.2).

During this analysis, several initial themes emerged, which were later refined and categorized. These initial themes included reflections on early ethical considerations and social implications for sexual health. The topic was addressed with a focus on understanding the nuanced ethical issues and social implications that affect sexual health in its early stages. In addition, a gender dimension was added, recognizing the importance of gender dynamics in articulating ethical aspects and health implications. Based on the recommendations of the 2030 Agenda for Sustainable Development, the aim was to emphasize equality, needs, distribution of resources, decision-making processes and power dynamics in the study of early ethical aspects and social impacts on sexual health.

FIGURE 12.2 As a result of delving into the abstract material, a mind map illustrating initial themes and subthemes emerged.

12.5 FINDINGS AND ELABORATE RESEARCH

12.5.1 CULTURAL AND SOCIAL FACTORS

Gender norms and roles: These are society's expectations and beliefs about the behaviors, roles and characteristics that are considered appropriate for individuals based on their gender. Gender norms and roles can vary from culture to culture and have a significant impact on many areas of life, including education, employment, relationships and healthcare.

Societal attitudes towards sexual health: This refers to the collective beliefs, perceptions and behaviors of the society on issues related to sex and health. The attitude of the society plays a vital role in how certain people are provided with the right treatment in the healthcare industry.

Reproductive rights and autonomy: These rights emphasize on the legal rights a citizen has over abortion, contraception, fertility treatments, maternal health and other health related services. Individuals must be able to have control and right over their own body and reproductive necessities.

12.5.2 ADVANCEMENTS IN THE FIELD OF TECHNOLOGY

The currently trending technologies like artificial intelligence (AI), machine learning (ML), and deep learning (DL) can help to bridge the data gap between different data available like genetic and epidemiological. We explore how these technologies can add depth to understand gender differences and develop effective interventions and progress in these areas:

1. **AI in Ethical standards for Genetic Testing:**
   - **Privacy:** Artificial Intelligence and ML algorithms can be used to analyze large genetic data structures for personalized medicine without compromising on human privacy. But there arises concern of re-identifying the genome structure and relating it to the individual, which states that data privacy should be highly effective.
   - **Explicit Consent:** AI-powered decision support systems can better explain to healthcare providers the complexities of genetic testing to patients by improvising informed consent processes. However, the ethical challenge is to ensure the genetic test results and potential risks are fully understood by the patients.
   - **Genetic Discrimination:** ML algorithms can be used to identify and classify patterns of genetic discrimination in healthcare and insurance settings, avoiding the risk of creating a discriminatory practice based on related genetic data.
   - **Data Secrecy:** AI provides encryption techniques to improve the security of storage and transmission of sensitive genetic information and ensure the data shared by the patient is safe. This gets tougher as maintaining confidentiality is increasingly difficult due to the huge genetic databases, requiring strict privacy measures.

- **Emotional Consequences:** ML algorithms can be taught to see well ahead the individuals' psychological responses to results from tests done on genetic testing results, making healthcare providers give customized tailored counseling and support services. Also providing family with sufficient support for genetic testing means.

12.5.3 AI IN E-HEALTH SOLUTIONS FOR GENDER WELLNESS

- **Online Telemedicine:** Telemedicine platforms can be developed with the help of chatbots and virtual assistants powered by AI by offering customized health information, scheduling appointments and support for gender-affirming care. But ethical concerns may arise concerning the accuracy of AI-driven medical advice and the need for human oversight to ensure empathetic and culturally competent care.
- **Mobile Health Apps:** User data can be analyzed by ML algorithms from mobile health apps to deliver personalized recommendations for gender-specific health interventions, such as hormone therapy management and mental health support. Ethical considerations include, ensuring the transparency of data usage and protecting user privacy.
- **AI-Based Online Platforms:** To track current trends and barriers in finding gender-affirming care, informing policy development and resource allocation, DL models can be trained to examine online discussions and social media interactions. Ethical concerns involve protection against algorithmic biases and shielding the anonymity of individuals sharing sensitive health information online.

In detail AI, ML, and DL technologies propose incredible potential to deal with ethical considerations in genetic testing and digital healthcare solutions for gender health. However, ethical oversight, transparency and the protection of individual rights and privacy remain supreme to ensure the responsible and equitable implementation of these technologies in healthcare.

12.5.4 INTERSECTIONALITY

Gender, ethnicity, and race health disparities: Race and ethnicity intersect with gender to create unique patterns Intersectionality recognizes that individuals hold multiple social identities, such as race, ethnicity, gender, sexuality and socioeconomic status, which intersect and interact to shape their experiences and health outcomes. Race and ethnicity intersect with gender to create unique patterns of health disparities, including disparities in access to healthcare, healthcare utilization, treatment outcomes, and health status.

Transgender healthcare disparities: Transgender individuals face significant healthcare disparities, including barriers to care, discrimination, and inequities in health outcomes. These disparities are influenced by factors such as stigma, discrimination, lack of culturally competent care, and legal barriers to accessing gender-affirming healthcare services.

12.5.5 SOCIOECONOMIC STATUS

Employment status and healthcare coverage: Socioeconomic factors, such as employment status, income level and access to health insurance have a profound impact on access to healthcare services and health outcomes. Employment status can determine whether individuals have access to employer-sponsored health insurance or must rely on public health insurance programs or out-of-pocket payments for healthcare.

Education level and health literacy: Education level and health literacy are critical determinants of health outcomes. Higher levels of education are associated with better health knowledge, healthier behaviors, increased healthcare utilization and improved health outcomes. Health literacy influences individuals' ability to understand health information, make informed decisions about their health and pilot the healthcare scheme efficiently.

12.5.6 HEALTHCARE SYSTEMS

Inclusion of diverse gender identities in healthcare: To provide equitable and culturally competent care, healthcare systems must be comprehensive of varied gender identities and expressions. This includes respecting individuals' gender identities, providing gender-affirming care, addressing the exclusive healthcare needs of transgender and gender nonconforming individuals and creating inclusive healthcare environments free from discrimination and bias.

Equitable access to healthcare services: Equitable access to healthcare services ensures equal opportunities to access timely, affordable and high-quality healthcare to all individuals, regardless of gender, race, ethnicity, socioeconomic status or other factors. Achieving equitable access requires addressing barriers such as geographic disparities, transportation issues, financial constraints, language barriers, discrimination and cultural differences.

12.5.7 AI IN LEGAL FRAMEWORKS

Gender-based discrimination laws: AI can analyze large datasets of Gender discrimination laws prohibit discrimination based on gender identity or expression in various domains, including employment, education, housing, healthcare and public accommodations. these Analysis can defend persons from discrimination, harassment and unequal treatment based on their gender uniqueness or expression.

Formulating reproductive rights legislation: AI classification algorithms can be used to forecast and inform right strategies to law formulating bodies related to access to contraception, abortion, prenatal care, maternal healthcare services and reproductive healthcare information. These laws can help protect individuals' rights to make decisions about their reproductive wellbeing and ensure safe access to comprehensive reproductive healthcare services, regardless of gender, race, ethnicity, socioeconomic status or other factors.

12.6 STRENGTHS AND LIMITATIONS

Searches were carried out in both Scopus and Web of Science, two expansive databases that collectively encompass over 20,000 and 12,000 journals, respectively during the data collection phase. Despite these efforts, there is a possibility that relevant articles were missed. The research process was governed by transparency, as evidenced by the thorough documentation of the methodological steps taken to develop the gender ethics model as shown in Table 12.1. This ensured an audit trail for readers, enhancing the dependability and validity of the results through a combination of gender theory, qualitative thematic analysis, and a modified systematic literature search. Noble and Heale's approach offers a more objective explanation.

Furthermore, using investigator triangulation, validity was improved. Three researchers with varied backgrounds comprised the research team, which enhanced the gender analysis. On the other hand it is critical to recognize that the understanding of texts can be influenced by the background and perspective of the researcher and that researchers who possess different viewpoints may present another interpretation. Research triangulation was implemented to address this issue and to enhance confidence and consistency, involving the complete involvement of the research team in the data.

Wide interdisciplinary experiences contributed to the development of the gender ethics model (GEM) as shown in Figure 12.3, with particular stress on the following thematic domains: technological advancements, intersectionality, socioeconomic

TABLE 12.1
Themes and Subthemes and Included Articles from the Search

Themes	Subthemes	Number of articles in each theme In total n = 61
Cultural and Social Factors	Gender norms and roles	24,25,26,27,28,29,30(n=7)
	Societal attitudes towards gender health	31,32,33(n=3)
	Reproductive rights and autonomy	34,35,36,37,38,39,40,41(n=8)
Technological Advancements	Ethical implications of genetic testing	1,2,4,6,7,46,47 (n=7)
	Digital healthcare solutions for gender health	
Intersectionality	Race, ethnicity, and gender health disparities	34,35,36,37,38,39,40,41 (n=8)
	LGBTQ+ healthcare disparities	
Socio Economic Status	Employment status and healthcare coverage	50, 24,25,26,27,28(n=6)
	Education level and health literacy	
Healthcare Systems	Inclusion of diverse gender identities in healthcare	45,46,47,48,49,50 (n=6)
	Equitable access to healthcare services	
Legal Frameworks	Gender discrimination laws	41,42,43,44,
	Reproductive rights legislation	45,46,47,48,49,50(n=10)
		51,14,15,23,37,27(n=6)

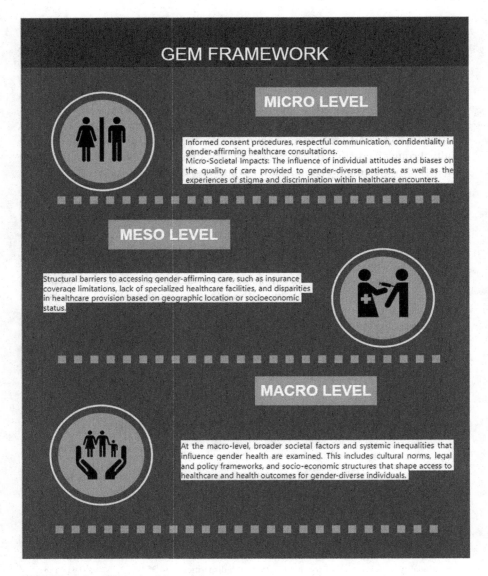

FIGURE 12.3   Gender health model.

status, healthcare systems, cultural and social factors, legal frameworks. The incorporation of these shared experiences improved comprehension and shaped the all-encompassing character of the model.

12.7   CONCLUSION

In undertaking this endeavor, a modified systematic literature review approach coupled with thematic analysis to develop a gender ethics framework was showcased. This framework aimed to integrate both individual and structural components, culminating in GEM. Gender mainstreaming is a strategy for integrating gender perspectives into all aspects of policy-making, programming, and organizational practices. The following are seven principles for effectively mainstreaming gender:

1. **Commitment from the Top Management**: The commitment of leadership is significant for successful gender mainstreaming. The clear commitment to gender equality from the senior management is necessary and it should ensure that gender equality is incorporated in all organizational activities.
2. **Integration into Policies and Practices**: Rather than considering gender equality as a separate issue, it should be a part in all policies, programs, and practices. This needs to be enacted by assessing the gender implications in current policies and practices followed by making necessary changes that will promote gender equality.
3. **Active Participation and Engagement**: It is the responsibility of all the stakeholders, including women and men from diverse backgrounds to actively participate in Gender mainstreaming. The engagement of affected communities in decision making processes is necessary and it is needed to consider their perspectives also.
4. **Gender Analysis and Data Collection**: To understand the requirements, priorities, and experiences of girls, boys, women and men, gender analysis should be conducted and sex-disaggregated data needs to be collected. Based on this information gender-responsive policies and programs are to be designed.
5. **Capacity Building and Training**: Enhancing the capacity of staff and stakeholders on gender equality and mainstreaming is required for effective implementation of gender-responsive approaches. Gender-sensitive initiatives and strategies, along with comprehensive gender analysis, are imperative in training programs to advance workplace gender equality. Integrating intersectional approaches and inclusive frameworks is essential for fostering equitable and transformative organizational environments.
6. **Resource Allocation and Budgeting**: Adequate resources are essential to support the efforts taken for gender mainstreaming. This can be done by including a specific budget for conducting gender-specific programs and activities. Integration of gender considerations in overall budgeting is also crucial.
7. **Monitoring and Evaluation**: The progress of gender mainstreaming initiatives should be monitored and evaluated regularly to ensure accountability and learning purposes. The impact of policies and programs on gender equality outcomes need to be monitored by regular assessments to find the areas of improvement.

By following these principles, policymakers and organizations can seamlessly incorporate gender perspectives on their endeavors and contribute to nurture gender equality and empowerment of women.

REFERENCES

Alang, S., McAlpine, D., McCreedy, E., and Hardeman, R. (2020). Police brutality and Black health: Setting the agenda for public health scholars. *American Journal of Public Health*, 110(5), 662–664.

American Medical Association. (2018). *AMA policy on transgender individuals*. American Medical Association. Retrieved from https://www.ama-assn.org/

Barrios, M., and Villarroya, A. (2022). What is needed to promote gender equality in the cultural sector? Responses from cultural professionals in Catalonia. *European Journal of Cultural Studies*, 25(4), 973–992.

Beauchamp, T. L., and Childress, J. F. (2013). *Principles of biomedical ethics* (7th ed.). Oxford University Press.

Beltran, A., and Puga, A. (2012). Paradigmatic changes in gender justice: The advancement of reproductive rights in international human rights law. Creighton International and Comparative Law Journal, 3, 158.

Beninger, C. (2021). Reproductive rights, UN sustainable development goals and international human rights law. *Gender Equality*, 1013–1025.

Bobbitt-Zeher, D. (2020). Gender discrimination policy. *Companion to Women's and Gender Studies*, 327–345.

Bond, J. E. (2004). Intersecting identities and human rights: The example of romani women's reproductive rights. *Georgetown Journal of Gender and the Law*, 5, 897.

Caldwell, J. C. (1990). Cultural and social factors influencing mortality levels in developing countries. *The Annals of the American Academy of Political and Social Science*, 510(1), 44–59.

Cook, R. J. (1991). International protection of women's reproductive rights. *New York University Journal of International Law and Politics*, 24, 645.

Correa, S., and Petchesky, R. (2007). Reproductive and sexual rights: A feminist perspective. *Culture, society and sexuality*. Routledge, 314–332.

Crenshaw, K. (1989). Demarginalizing the intersection of race and sex: A black feminist critique of antidiscrimination doctrine, feminist theory and antiracist politics. *University of Chicago Legal Forum*, 1989(1), 139–167.

Escobar, J. I., and Gureje, O. (2007). Influence of cultural and social factors on the epidemiology of idiopathic somatic complaints and syndromes. *Psychosomatic Medicine*, 69(9), 841–845.

Falk, A., and Hermle, J. (2018). Relationship of gender differences in preferences to economic development and gender equality. *Science*, 362(6412), eaas9899.

Green, A. R., Betancourt, J. R., and Emilio Carrillo, J. (2002). Integrating social factors into cross-cultural medical education. *Academic Medicine*, 77(3), 193–197.

Grown, C., Gupta, G. R., and Pande, R. (2005). Taking action to improve women's health through gender equality and women's empowerment. *The lancet*, 365(9458), 541–543.

Hankivsky, O. (2012). Gender, sex, and health equity: Intersectional approaches to health disparities. *Health Sociology Review*, 21(1), 1–7.

Hayden, L. A. (1998). Gender discrimination within the reproductive health care system: Viagara v. birth control. *JL & Health*, 13, 171.

Hendriks, A. (1994). Promotion and protection of women's right to sexual and reproductive health under international law: The economic covenant and the women's convention. *American University Law Review*, 44, 1123.

Howell, E. A., Egorova, N., Balbierz, A., Zeitlin, J., and Hebert, P. L. (2016). Site of delivery contribution to black-white severe maternal morbidity disparity. *American Journal of Obstetrics & Gynecology*, 215(2), 143–152.

Htun, M. (2004). Rising tide: Gender equality and cultural change around the world. *Comparative Political Studies*, 37, 738.

Institute of Medicine. (2001). *Crossing the quality chasm: A new health system for the 21st century* (pp. 1–337). National Academy Press.

Jain, D., and Shah, P. K. (2020). Reimagining reproductive rights jurisprudence in India: Reflections on the recent decisions on privacy and gender equality from the supreme court of India. *Columbia Journal of Gender and Law*, 39, 1.

James, S. E., Herman, J. L., Rankin, S., Keisling, M., Mottet, L., and Anafi, M. (2016). The report of the 2015 U.S. transgender survey. National Center for Transgender Equality.

Johnson, M. J., Nemeth, L. S., Mueller, M., Eliason, M. J., Stuart, G. W., and Johnson, T. R. B. (2014). On the outside looking in: Promoting inclusion in healthcare organizations. *Journal of Public Health Management and Practice*, 20(6), E1–E9.

Krieger, N. (2012). Methods for the scientific study of discrimination and health: An ecosocial approach. *American Journal of Public Health*, 102(5), 936–944.

Kuhlmann, A. S., Curran, L., and VanDevanter, N. (2017). "You just have to adapt": LGBT+ young adults and the negotiation of healthcare. *Journal of Homosexuality*, 64(10), 1397–1417.

Loll, D., et al. (2021). Factors associated with reproductive autonomy in Ghana. *Culture, Health & Sexuality*, 23(3), 349–366.

Lomazzi, V. (2022). Gender equality values and cultural orientations. *Reflections on European Values: Honouring Loek Halman's Contribution to the European Values Study*.

Lombardi, E. L., Wilchins, R. A., Priesing, D., and Malouf, D. (2001). Gender violence: Transgender experiences with violence and discrimination. *Journal of Homosexuality*, 42(1), 89–101.

MacKinnon, C. A. (1991). Reflections on sex equality under law. *Yale Law Journal*, 1281–1328.

McDougall, R. J. (2020a). Moral complexities in prenatal screening and gender-affirming treatments. *Bioethics*, 34(3), 215–222.

McDougall, R. (2020b). The ethics of gender-affirming care for transgender youth. *AMA Journal of Ethics*, 22(1), E66–E72.

Nepal, A., et al. (2023). Factors that determine women's autonomy to make decisions about sexual and reproductive health and rights in Nepal: A cross-sectional study. *PLOS Global Public Health*, 3(1), e0000832.

Obermeyer, C. M. (1995). A cross-cultural perspective on reproductive rights. *Human Rights Quarterly*, 17, 366.

O'Sullivan, M. (2007). Reproductive rights. *Constitutional Law of South Africa(Juta Cape Town Revised Service 3 2011)*, 37-1.

Pacelle, R. L. (2003). *Between law and politics: The solicitor general and the structuring of race, gender, and reproductive rights litigation*. No. 14. Texas A&M University Press.

Payne, S. (2015). The health of women and girls: How can we address gender equality and gender equity? *Seminars in Reproductive Medicine*, 33(1). Thieme Medical Publishers.

Pillard, C. T. L. (2006). Our other reproductive choices: Equality in sex education, contraceptive access, and work-family policy. *Emory Law Journal*, 56, 941.

Princewill, C. W., et al. (2017). Autonomy and reproductive rights of married Ikwerre women in Rivers State, Nigeria. *Journal of Bioethical Inquiry*, 14, 205–215.

Reisner, S. L., Poteat, T., Keatley, J., Cabral, M., Mothopeng, T., Dunham, E., . . . Baral, S. D. (2016). Global health burden and needs of transgender populations: A review. *The Lancet*, 388(10042), 412–436.

Shalev, C. (2000). Rights to sexual and reproductive health: The ICPD and the convention on the elimination of all forms of discrimination against women. *Health and Human Rights*, 38–66.

Shaw, A. (2018). Gender and resource allocation: A policy perspective. *Journal of Social Policy*, 45(3), 345–360.

Siddiqi, N., and Shafiq, M. (2017). Cultural value orientation and gender equity: A review. *Social Psychology &Society*, 8(3).

Stark, B. (2010). The women's convention, reproductive rights, and the reproduction of gender. *Duke Journal of Gender Law & Policy*, 18, 261.

Tadiri, C. P., et al. (2021). Determinants of perceived health and unmet healthcare needs in universal healthcare systems with high gender equality. *BMC Public Health*, 21, 1–9.

Tesch-Römer, C., Motel-Klingebiel, A., and Tomasik, M. J. (2008). Gender differences in subjective well-being: Comparing societies with respect to gender equality. *Social Indicators Research*, 85, 329–349.

Thein, P. T. (2015). Gender equality and cultural norms in Myanmar. *International Conference on Burma/Myanmar Studies (Jul. 2015)*.

Thi, H. D., et al. (2023). Socio-cultural norms and gender equality of ethnic minorities in Vietnam. *Journal of Racial and Ethnic Health Disparities*, 10(5), 2136–2144.

Tong, R. (1993). Feminist ethics. In E. N. Zalta (Ed.), *The stanford encyclopedia of philosophy* (Winter 2016 Edition). Retrieved from https://plato.stanford.edu/archives/win2016/entries/feminism-ethics/

Uberoi, D., and De Bruyn, M. (2013). Human rights versus legal control over women's reproductive self-determination. Health and Human Rights, 15, 161.

Varnum, M. E. W., and Grossmann, I. (2016). Pathogen prevalence is associated with cultural changes in gender equality. *Nature Human Behaviour*, 1(1), 0003.

Wang, G.-Z. (2016). *Reproductive health and gender equality: Method, measurement, and implications*. Routledge.

Whitehead, J., Shaver, J., and Stephenson, R. (2016). Outness, stigma, and primary health care utilization among rural LGBT populations. *PLoS ONE*, 11(1), e0146139.

Williams, D. R., and Mohammed, S. A. (2013). Racism and health I: Pathways and scientific evidence. *American Behavioral Scientist*, 57(8), 1152–1173.

World Health Organization. (2019). Transgender health: Taking stock of the evidence and setting an agenda for change. Retrieved from https://apps.who.int/iris/bitstream/handle/10665/311940/WHO-HIV-2019.02-eng.pdf

13 Examination of AI's Role in Diagnosis, Treatment, and Patient Care

Seyed Mahmoud Sajjadi Mohammadabadi,
Faezehalsadat Seyedkhamoushi, Mehrnaz
Mostafavi, and Mahsa Borhani Peikani

13.1 INTRODUCTION

The healthcare industry faces ever-growing demands, requiring innovative solutions. AI and ML are emerging as powerful tools with the potential to revolutionize patient care. These technologies can transform every aspect of the care journey, from diagnosis and treatment planning to ongoing support and monitoring.

This transformation is particularly timely given the immense pressure healthcare workers faced during the COVID-19 pandemic. Studies have shown significantly higher burnout and stress among those directly caring for patients compared to non-patient-facing staff [1]. Another study shows providing adequate personal protective equipment can significantly reduce the stress levels of healthcare workers [2]. Also, researchers analyzed stress within a hospital and discovered that medical staff generally experienced less stress than non-medical staff. However, workload remained the most significant source of stress for all employees [3]. The healthcare system is experiencing a significant transformation caused by the integration of AI and ML. AI is promising in healthcare, as it revolutionizes patient care from diagnosis to support. AI algorithms are used to analyze big data of medical data, including patient records, imaging results, and genomic information. This analysis allows researchers to identify patterns and trends that might not be easily discovered by human observation. AI also can leverage medical data to predict potential health risks and tailor treatment plans to individual patients [4], which is called personalized medicine. Healthcare specialists can make real-time insights and recommendations using clinical decision support using AI. These systems suggest the most effective treatment by analyzing a patient's medical history, current condition, and available treatment options. AI has the potential to improve remote patient monitoring and proactive healthcare. By analyzing data collected from wearable devices and sensors, these systems can continuously monitor a patient's health. This monitoring enables early intervention and preventative measures. Therefore leads to improved patient health and gives a more proactive approach [5].

DOI: 10.1201/9781003473435-13

While AI and ML offer exciting possibilities for the future of healthcare, it's important to acknowledge the valuable role traditional diagnostic techniques continue to play. For instance, a widely used staining technique in biology allows for the evaluation of a large number of patient samples during interventional procedures. This technique, readily available as a standard service in pathology labs [6], enables the analysis of tissue samples to reveal various inflammatory features, including multinucleated cells, immune cells, and cellular inclusions [7].

13.2   ARTIFICIAL INTELLIGENCE IN HEALTHCARE DELIVERY

Machine learning, a powerful branch of AI, is transforming healthcare. It enables the analysis of vast medical data to uncover hidden patterns, ultimately improving disease detection and personalizing treatment plans. Figure 13.1 illustrates the two main approaches in AI: identification and classification.

13.2.1   PATTERNS WITH MACHINE LEARNING FOR HEALTHCARE

Machine learning allows computers to learn from data, uncovering patterns relevant to healthcare, including those related to gender. Table 13.1 provides a summary of several commonly utilized AI algorithms

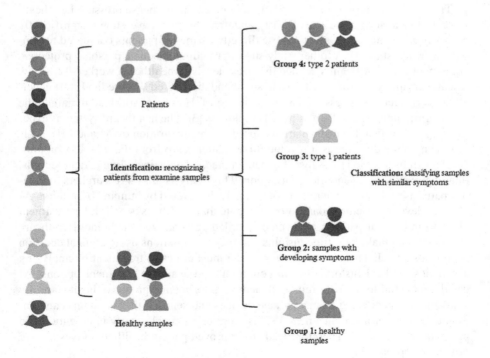

FIGURE 13.1   Identification and classification of patients using machine learning.

TABLE 13.1
Common AI Algorithms and Their Characteristics

Algorithm	Description	Advantages	Limitations
Artificial Neural Network (ANN)	Interconnected network processing information through layers.	Highly flexible, learns complex relationships, effective for various tasks.	Computationally expensive, challenging to interpret, requires large training datasets.
Decision Tree	Treelike structure representing data relationships.	Easy interpretation, handles mixed data types.	Prone to overfitting, susceptible to data noise.
Logistic Regression	Performs binary classification with linear relationship.	Simple interpretation, efficient for small datasets.	Prone to overfitting complex data, limited to binary classification.
LASSO Regression	Regularized linear regression with penalty term.	Handles high-Dimensional data, Improves inter-pretability.	Requires tuning, computationally expensive.
Support Vector Machine (SVM)	Finds optimal hyper-plane to separate data into classes.	Robust to outliers, Effective for high-dimensional data.	Can be computationally expensive, complex interpretation.
Random Forest	Ensembles multiple decision trees for accuracy.	Robust to outliers, handles high-dimensional data.	Can be computationally expensive, with less interpretability.
Naive Bayes	Uses Bayes' theorem with assumed variable independence.	Efficient for small datasets, multiclass capable.	Independence as-Sumption may be violated, noise sensitivity.
K-Nearest Neighbors (kNN)	Predicts class or value based on nearest neighbors.	Simple implementation, effective for nonlinear relations.	Sensitive to data dimensionality, outliers, computational expense.

13.2.2 DATA PRIVACY AND SECURITY IN AI-POWERED HEALTHCARE

Recent developments in AI and ML, such as generative AI, are demonstrating promising AI applications in both industry and healthcare [8], [9]. These advancements are continuously improving data analysis and model development across various fields [10]. The growing adoption of large language models (LLMs) in healthcare raises crucial questions about patient data privacy and security. Researchers are actively exploring techniques to train LLMs while safeguarding privacy. One promising approach is federated learning (FL). FL enables collaborative model development without compromising patient data confidentiality, as the training data remains on individual devices [11]. However, implementing FL for complex, large-scale models present a significant hurdle: the vast differences in computational and communication capabilities across devices. Additionally, the size and complexity of healthcare data tasks can further complicate the process. To overcome these challenges associated with resource heterogeneity, researchers are developing advanced machine learning algorithms. These algorithms dynamically shift computationally demanding portions of the model to a central server, thereby reducing resource constraints on individual devices and accelerating the training process [12].

13.2.3 MACHINE LEARNING

Machine learning is transforming healthcare, enabling personalized medicine by analyzing vast amounts of patient data to identify gender-specific patterns and trends. This empowers healthcare professionals to:

- **Improve Diagnosis:** ML algorithms can analyze medical images, genetic data, and patient history to identify subtle differences in disease presentation between genders, leading to more accurate diagnoses.
- **Develop Personalized Treatment Plans:** By understanding how different genders respond to medications and therapies, ML can guide the development of personalized treatment plans that optimize efficacy and minimize side effects.
- **Predict Health Risks:** ML models can analyze various factors, including genetics, lifestyle choices, and social determinants of health, to predict health risks specific to different genders. This allows for early intervention and preventive measures.

Machine learning empowers computers to learn from vast healthcare datasets, uncovering patterns relevant to gender. This allows for a more personalized approach to healthcare compared to traditional one-size-fits-all methods.

13.2.4 AI APPROACHES AND THEIR APPLICATIONS IN HEALTHCARE

Training ML algorithms involves various approaches, with supervised, unsupervised, and reinforcement learning being the most common:

Supervised Learning: This training involves training algorithms on labeled datasets to make predictions for unseen and unlabeled data. For example, supervised learning algorithms can be used to analyze medical images like X-rays or mammograms to identify potential abnormalities. Supervised Learning is commonly used in healthcare studies.

Unsupervised Learning: This approach is used to find patterns and hidden structures in unlabeled data. Unsupervised learning group patients with similar characteristics.

Reinforcement Learning: This approach allows machines to learn through trial and error to achieve specific goals. It has potential applications in areas like robotic-assisted surgery, where robots can learn and refine their movements based on feedback from surgeons.

Understanding these different approaches is crucial for healthcare professionals working with ML. When selecting an ML model for a specific application, factors like data availability, model interpretability, and regulatory considerations should be taken into account.

This book, however, focuses on how AI and ML can transform gender-based healthcare. Traditionally, medical research and treatment approaches haven't always considered the unique biological and social factors affecting men and women differently. AI, with its ability to analyze vast amounts of patient data, holds immense promise for developing gender-specific treatment plans and improving healthcare outcomes for all.

13.2.5 DIFFERENT DATA TYPE

The growing application of ML in analyzing data related to rheumatic autoimmune inflammatory diseases (RAIDs) is being explored [13]. ML, which recognizes patterns and relationships in data, is being used to classify patients with RAIDs based on various types of data such as medical records, imaging, genetics, and transcriptomics. However, many studies in healthcare systems face some limitations such as small sample sizes and inaccurate sample labels. Also, researchers should be careful about overfitting and underfitting ML models, which can reduce the generalization of the model. Despite challenges, ML has the potential to aid in precision medicine and improving patient care.

13.2.5.1 Different Data Types for AI in Healthcare

To have efficient ML in healthcare, the data used to train it needs to be accurate and come from many different sources. These ML programs can then analyze this information to find patterns, make predictions, and help physicians.

The following are the different data types in healthcare:

Electronic Health Records (EHRs): Usually, hospitals and clinics store information about their patients electronically, like age, past health problems, medications, and test results. ML systems can analyze this information to find patients at risk for certain diseases, predict their disease progress, and create personalized treatment plans. For example, they can look at a patient's family history, blood pressure, and cholesterol to see their risk of heart disease. This information can also help physicians avoid giving patients medications that might interact poorly. By using this approach, healthcare system can be improved.

Medical Imaging: Images such as X-rays, MRIs, CT scans, and mammograms can be used by researchers and physicians to visualize the inside of a patient's body. These images are essential for identifying health problems, accurately diagnosing diseases, and monitoring treatment effectiveness. While regular X-rays remain valuable in healthcare systems, newer approaches like MRIs are better suited for detecting diseases in their early stages. AI algorithms can complement traditional medicine in various medical fields such as eye care and skin care. For instance, AI algorithms can analyze MRI scans to identify early signs of brain disorders like Alzheimer's disease, enabling earlier treatment initiation and enhancing patient health. Another example of AI advancement is the analysis of eye scans to detect diabetic retinopathy, a leading cause of blindness.

Furthermore, ML methods can be developed to analyze skin lesions to determine if they might be melanoma (skin cancer).

Genomic Data: The human genome holds valuable information about our health. By analyzing an individual's unique genetic data, patient-specific genomic data, ML algorithms can be used to gain deeper insights into human health. Different ML algorithms have been developed to analyze this data, enabling the identification of patients at high risk for specific diseases. Based on that analysis, ML enables personalized treatment tailored to the unique genetic profile of each patient. As an instance, genomic analysis may predict cancers, allowing for early intervention and the preparation of personalized treatment plans. Additionally, genomic data can identify patients who may experience favorable responses to specific medications.

Biomedical Signals: Biomedical signals, including heart rate, blood pressure, and electroencephalogram (EEG) data, provide real-time information about a patient's health. By analyzing these signals, we can detect early signs of illness, and monitor chronic conditions. As an example, AI can analyze blood sugar levels and identify patterns that predict potential diseases.

Biomedical signals, such as heart rate, blood pressure, and EEG data, offer real-time insights into a patient's health status. By analysing these signals, physicians can detect early signs of illness and continuously monitor chronic conditions. For instance, blood sugar levels are fed to ML algorithms, which identify patterns that forecast diabetes and other potential diseases.

Wearable Device Data: Wearable trackers have become increasingly popular and nowadays are used by many. These devices provide us with information about our activity, sleep, and heart rate etc. Information from wearable trackers can be used in preventing health issues. For example, they can alert users about their long periods of inactivity, which can be unhealthy. Also, by tracking sleep quality and heart rate, these devices provide more insights into stress levels of individual.

While gathering a large amounts of healthcare data presents opportunities for AI, some limitations remain. Ensuring data quality, protecting patient privacy, and integrating information from various sources requires careful consideration when developing AI applications. Considering these challenges is critical in transforming healthcare.

Table 13.2 summarizes the advantages and challenges of various data types in healthcare systems.

13.2.6 AVAILABLE DATASETS FOR RESEARCH

Public dataset helps researchers develop AI models for healthcare. These datasets are necessary for researchers to train, and test ML models. Table 13.3 summarizes public datasets for healthcare.

1. **NIH Chest X-Ray Dataset:** This dataset consists of chest X-ray images, labeled by pathologies, which can be used to train AI models for chest disease classification.

TABLE 13.2
Advantages and Challenges of Different Data Types

Data Type	Advantages	Challenges
Electronic Health Records (EHRs)	Rich source of patient demographics, medical history, diagnoses, medications, and lab results. Enables identification of disease risks, prediction of patient outcomes, and personalization of treatment plans.	Data quality and accuracy can vary. Manual data extraction can be time-consuming and error prone. Integrating data from disparate EHR systems requires standardization.
Medical Imaging (X-ray, MRI, CT scan, etc.)	Provides valuable insights into patient anatomy and physiology. Enables detection of abnormalities, supports accurate diagnosis, and monitors treatment progress.	Traditional modalities may have limitations in early disease detection. Requires expertise for image interpretation.
Genomic Data	Provides a patient's genetic makeup. Enables identification of individuals at risk for specific diseases and personalized treatment approaches.	Ethical considerations regarding data privacy and ownership. Requires specialized expertise for data analysis and interpretation.
Biomedical Signals (heart rate, blood pressure, EEG)	Provides real-time information about a patient's health status. Enables early detection of illness, monitoring of chronic conditions, and personalized interventions.	Requires reliable and secure wearable devices for data collection. Data quality can be affected by factors like sensor placement and patient behavior.
Wearable Device Data (activity levels, sleep patterns, vital signs)	Enables continuous monitoring of health parameters and promotes preventive healthcare. Provides personalized wellness recommendations based on individual data.	Data security and privacy concerns regarding continuous data collection. Potential for data bias based on user demographics and device limitations.

2. **Women's Health Initiative (WHI) Study:** This dataset of women's health across various life stages presents a dataset for researchers investigating gender-specific health issues.

3. **UK Biobank:** This population-based dataset includes genetic data, clinical data, and lifestyle factors from over 500,000 participants. Researchers can leverage this data to explore the interplay between genetics, gender, and disease susceptibility. Understanding how gender interacts with genetic variations can lead to the development of more personalized and gender-tailored healthcare strategies.

4. **CDC National Health and Nutrition Examination Survey (NHANES):** This survey provides a wealth of data on the health and nutritional status of the US population, including information on demographics and gender. Researchers can utilize NHANES data to identify and address potential gender disparities in healthcare access, disease prevalence, and treatment outcomes.

TABLE 13.3
Public Datasets for AI in Gender-Inclusive Healthcare Research

Dataset	Type	Size	Focus
NIH Chest Xray dataset	Medical Imaging (Xray)	112,120 images	Chest pathology classification (pneumonia, cardiomegaly, etc.)
Women's Health Initiative (WHI) Study	Clinical data, lifestyle factors	Over 68,000 women	Women's health across various life stages
UK Biobank	Genomic data, clinical data, lifestyle factors	Over 500,000 participants	Population-based health information, including genetic data
CDC National Health and Nutrition Examination Survey (NHANES)	Clinical data, demographic data	134,310 participants	National health and nutrition data in the US population
MIMIC (Multiparameter Intelligent Monitoring in Intensive Care) Database	Clinical data	112,000 clinical reports	Intensive care unit (ICU) patient data for critical care research
Optum deidentified Database	Electronic health records (EHRs)	Over 100 million patients	De-identified EHR data from a large US healthcare system

5. **MIMIC Database:** This ICU patient data repository can be used to develop AI models that improve critical care delivery, considering potential gender differences. Research suggests that men and women may respond differently to critical care interventions, necessitating gender aware approaches in AI-powered ICU management.

6. **Optum De-Identified Database:** This large collection of de-identified EHR data can be a valuable resource for researchers investigating gender bias in healthcare diagnoses and treatment patterns.

The selection of an appropriate dataset hinges on your specific research objectives and focus area. Factors such as data type (images, clinical data), size, and target variable should all be considered when making your choice.

13.2.7 Cloud Computing: Powering AI in Healthcare

The research by [14] explores how big data and AI can revolutionize healthcare. By analyzing data from different sources, such as medical records and wearable data, AI has potential for identifying disease and their risk, personalizing treatment, and enabling early detection of illnesses. However, some challenges related to data quality and privacy must be considered to fully leverage ML potential.

Cloud computing is a critical infrastructure for developing AI algorithms in healthcare. Managing massive datasets gathered by various healthcare systems is important for using AI algorithms. In the following, we discuss how cloud computing is necessary for using machine learning for various health-care applications:

1. **Scalable Data Management:** Traditional computing systems installed within healthcare facilities are not suitable to manage and analyze the growing medical data, such as medical images, EHRs, and genomic information. Cloud platforms are recent solutions for this outdated infrastructure by providing flexible storage and computing resources that can expand as needed. This simplifies the storing, processing, and analysis of big health-care datasets. allowing researchers and healthcare providers to benefit from AI models effectively. As a result, cloud computing is more reliable and ultimately benefits more patients.

2. **Collaborative Research:** Cloud platforms become hubs for research collaboration. These platforms offer a common environment where researchers and clinicians from different institutions access and analyze data collectively. This shared space enables data sharing, collaborative AI model development, and fast validation of research discoveries.

3. **Resource Allocation:** Cloud computing allows researchers and healthcare professionals to access computing resources as needed. They can scale their AI projects up or down according to requirements. This flexible approach minimizes expenses and maximizes resource efficiency, particularly beneficial for demanding tasks such as deep learning model training.

4. **Robust Security and Privacy:** Given the sensitivity of patient data, ensuring its security is critical. Cloud platforms offer many security features to comply with regulations. These measures protect patient privacy and confidentiality while allowing data utilization for research.

The combination of AI and cloud computing represents an opportunity to improve healthcare systems. Through data analysis, tailored treatment, and enhanced medical research.

13.2.8 Applications of Artificial Intelligence in Healthcare

AI is reshaping healthcare, with its potential to improve every aspect of diagnosis, treatment, and patient well-being. This section delves into some of the key domains where AI is developing. Figure 13.2 summarizes different applications of AI in healthcare.

1. **Disease Diagnosis and Risk Prediction:** AI algorithms can analyze medical data, including EHRs, medical imaging, and genetic data. Then they can be used to identify patterns and predict an individual's risk of developing certain diseases.

2. **Image Analysis and Feature Extraction:** AI models process and analyze medical images. They can automatically detect abnormalities, segment specific regions of data, and extract quantitative features that would be time-consuming for human analysis.

3. **Personalized Care:** AI models can assess a patient's medical background, present state, and how they've responded to past therapies to suggest personalized treatment. This method improves treatment effectiveness and minimizes adverse effects.
4. **Drug Discovery:** AI is enhancing drug discovery by examining extensive datasets containing molecular structures and patient information. This speeds up the discovery of potential drug candidates.
5. **Clinical Decision Support:** Clinical decision support systems, by using AI, offer real-time guidance and recommendations to healthcare. By analyzing patient data, these systems offer relevant treatment suggestions or diagnostic tests, enhancing clinical decision-making.
6. **Patient Monitoring and Early Detection:** ML algorithms can process data collected from wearable devices and sensors, allowing continuous monitoring of a patient's health status. This proactive approach enables early detection of potential health concerns.
7. **Chatbots and Virtual Assistants:** AI-driven chatbots and virtual assistants offer 24-hour access to information for patients, addressing basic healthcare questions and assisting with appointment scheduling.

This system encourages individuals to play a more proactive role in managing their health.

13.3 AI IN TRANSFORMING DISEASE DIAGNOSIS, DETECTION AND IDENTIFICATION

13.3.1 Enhancing Objectivity and Early Detection with AI

Accurate and timely disease detection is essential for effective healthcare. Traditional diagnosis methods can be inconsistent. AI presents solutions with its objective and

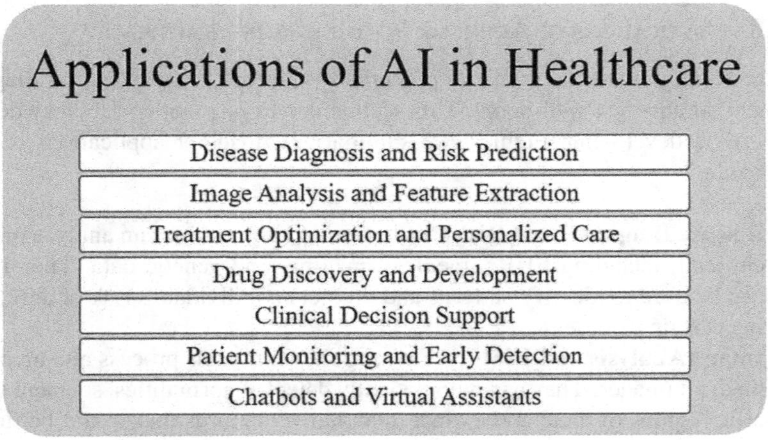

FIGURE 13.2 Applications of artificial intelligence in transforming healthcare.

automated diagnostic approaches. AI excels in analyzing extensive medical images and patient data, leveraging machine learning algorithms to detect patterns and abnormalities. This capability enables earlier and more precise diagnoses in the early stages of diseases.

Consider rheumatoid arthritis (RA), an autoimmune condition primarily affecting joints. Traditionally, assessing RA severity involves symptom evaluations and X-ray analysis. However, researchers explore machine learning methods for analyzing ultrasound images, offering more objective severity assessment. For instance, a study introduced an automated grading system using image processing and ML algorithms to segment bone erosion and synovial thickening, key severity indicators [15]. This method outperformed conventional approaches. By extracting quantitative features from segments and combining geometric and texture data, ML models achieved a 92.5% classification accuracy. This demonstrates AI's potential in diagnosis, providing more precise and consistent assessments for personalized treatment planning, regardless of gender.

13.3.2 EARLY DISEASE PREDICTION

Another significant application of AI is early disease prediction. ML algorithms can identify individuals at higher risk of developing a specific disease, by analyzing patient data, including medical history, blood tests, and genetic information. For instance, a study examined ML algorithms to predict chronic disease at its early stages [16].

13.3.3 GENDER-NEUTRAL AND PERSONALIZED MEDICINE

By analyzing a patient's medical and familial history, ML models can forecast an individual's disease vulnerability. This enables tailored medications. Moreover, AI has the potential to mitigate healthcare distinctions. When diagnostic tools might be less effective for a particular gender, AI can improve the procedure. ML models can learn to recognize disease patterns independent of gender for gender-neutral diagnostic solutions.

AI disease detection has many advantages, especially regarding gender considerations in healthcare:

- Integration into Clinical Workflows: Providing healthcare providers with real-time insights and decision-support tools that consider gender-specific elements during patient care.
- Discovery of New Biomarkers: Identifying new biomarkers, potentially leading to earlier and more accurate diagnoses, which can lead to the development of new targeted therapies that consider sex-based biological differences.

13.4 MACHINE LEARNING IN DISEASE CLASSIFICATION

Machine learning is revealing new approaches to disease diagnosis and classification. ML shows important features and classifies diseases alongside existing diagnostic methods. For example, a study checked if combining thermal imaging with

ML could help classify diseases [17]. They used decision tree and support vector machine (SVM) algorithms to increase accuracy. They measured metrics like accuracy, sensitivity, and ROC curves to see how well the models did. While the study showed this ML could work, we need more research to be sure it's reliable enough for real-world use. Another study focused on using feature selection and ML to classify different types of a specific disease [18]. They collected data from patients, selected the most important elements, and then tested different ML models. They found that ANN worked best after features engineering.

13.5 PRECISION MEDICINE AND TREATMENT RESPONSE

Healthcare is moving towards precision medicine. By considering genetics, environment, lifestyle, and specific biomarkers, precision medicine aims to optimize diagnosis, improve treatments, and minimize side effects.

A recent study examines ML to forecast sustained remission in patients with chronic diseases experiencing biological drug treatment [19]. ML algorithms can be employed to predict responses to different treatment options using omics data and imaging information [20]. Recent research explored ML's capability to predict a vision-threatening difficulty in patients with chronic disease [21]. Another study investigated ML models' capability to predict discontinuation of biosimilar drugs in patients with diseases [22].

While promising results have been achieved using ML to predict response to methotrexate (MTX), a treatment for RA, challenges continue [23]. These include an MTX-focused approach and limited research on cost-effective alternatives for other medications. However, recent research on ML predicting a vision-threatening complication in another disease indicates broader ML application potential [24]. Similarly, exploring cost-effective alternatives becomes imperative to ensure fair access to personalized medicine approaches across diverse patient populations [25].

In summary, while physicians traditionally relied on specific scoring systems for disease severity assessment, recent research shows ML algorithms outperforming these methods in predicting disease activity. Additionally, ML can analyze data from various imaging modalities to assess the risk of common comorbidities associated with the primary disease [26]. This capability extends beyond the initial disease, as ML can analyze data from different sources to assess the risk of complications patients are more susceptible to [26]. Wearable devices with specific functionalities further enhance disease monitoring and potentially predict future disease course [27].

Machine learning streamlines drug discovery for chronic diseases like rheumatism. Researcher use patient data to predict effective treatments, skipping initial guesswork and efficiently handling large datasets [28]. This data-driven approach, validated by external databases, holds promise for faster discovery of effective therapies.

13.6 GENDER-INCLUSIVE PATIENT CARE WITH
AI-POWERED TREATMENTS

Traditional healthcare often employed a "one-size-fits-all" approach to treatment plans, neglecting individual variations, especially those linked to gender. This approach can be ineffective. AI offers a chance to revolutionize healthcare by enabling the development of personalized treatment plans that consider individual characteristics, including gender.

Recent research emphasizes the integration of AI and precision medicine to enhance healthcare delivery [29]. AI algorithms are used to analyze big healthcare datasets. They provide insights that can significantly improve decision-making and patient health. The importance of considering sex and gender diversity in AI for medicine is emphasized by [30]. This research highlights how AI systems often ignore sex and gender identities, leading to treatment biases that can result in misdiagnosis, especially for women. Thus, managing possible biases in ML algorithms, as debated in [31], is critical.

AI suggests several methods to tailor personalized medicine based on gender:

- **Genomic Data Analysis:** By analyzing a patient's genomic data, healthcare professionals can identify patterns that may influence their response to medications. This analysis gives the capability to personalize medicine, allowing for the customization of drug selection and dosage, ultimately optimizing treatment effectiveness while minimizing side effects.
- **Treatment Outcomes:** AI-powered prediction empowers healthcare providers to make informed treatment decisions, by forecasting how patients may respond to various treatment options.

Despite these potential advantages, some challenges remain in implementing AI for personalized treatment:

- **Transparency and Explainability:** Comprehending the explanation behind AI-generated treatment is critical for fostering a trustworthy and ethical approach in healthcare.
- **Human Oversight:** AI should complement human expertise rather than replace it, particularly in considering complex gender-specific factors in personalized medicine.

Some researchers believe couples should have access to a procedure called PGD to choose the sex of their baby during IVF. They express that couples should have the right to make these choices, as long as the procedure is safe. PGD has a low risk of complications, and if couples know the pros and cons, then using it for gender selection might not be unethical [32].

13.7 ARTIFICIAL INTELLIGENCE LIMITATIONS AND CHALLENGES

AI in healthcare is a powerful tool, but it can also broaden existing gaps in care because of biased data, and how physicians use it. Fairness goes beyond race and considers social determinants of health. Techniques exist to mitigate bias, but challenges include dataset shift and limited data diversity. Ethical use of ancestry information is crucial [31].

AI is rapidly making strides in disease diagnostics. Research has exploded since 2013, with neural networks being the leading AI approach, particularly for cardiovascular diseases. However, the field is far from mature. Most studies focus on developing specific AI models for a single disease, with limited comparison between methods. Additionally, there's little research on integrating AI into real-world clinical settings. Furthermore, data quality and bias in training data are concerns. Researcher emphasizes AI's potential for diagnostics but highlights the need for further development [33]. Future research should focus on comparing AI methods, ensuring high-quality data, mitigating bias, and creating user-friendly interfaces for doctors to seamlessly integrate AI into their workflow.

13.7.1 DATA BIAS AND FAIRNESS

One of the most critical challenges facing AI in healthcare is data bias. AI models are trained on vast datasets, and the quality and representativeness of this data significantly impact their performance.

13.7.1.1 Transparency and Explainability

Another limitation is that some AI algorithms are like black boxes—you give them data and they give you an answer, but you can't see how they reached that answer.

13.8 ETHICAL CONSIDERATIONS

With AI changing healthcare, it's essential to consider the ethical side of using it. Some key concerns are keeping patient information private, making sure AI is fair, and ensuring ML models are transparent. While AI is promising, there's no replacing a physician's expertise. Think of AI as a fancy tool that helps physicians make better decisions, not take over their jobs. Physicians should always be in charge of diagnosis and treatment, using AI insights alongside their own judgment and years of experience. Collaboration between healthcare professionals and AI systems fosters a human-in-the-loop approach, ensuring responsible and ethical use of AI in patient care.

13.9 FUTURE DIRECTIONS

The transformative potential of AI in healthcare extends far beyond the current applications discussed in this chapter (see Figure 13.3). As research progresses and technology evolves, we can expect to see significant advancements in several key areas.

- Enhanced Clinical Workflows: Seamless integration of AI tools into clinical workflows will empower healthcare professionals with real-time insights and decision support systems tailored to consider gender-specific factors at the point of care. This can lead to more efficient and effective healthcare delivery, improving patient outcomes.
- Improved Diagnostic Accuracy: AI-powered diagnostics hold immense promise for achieving greater accuracy and consistency in disease detection and classification, regardless of a patient's gender. Advanced algorithms

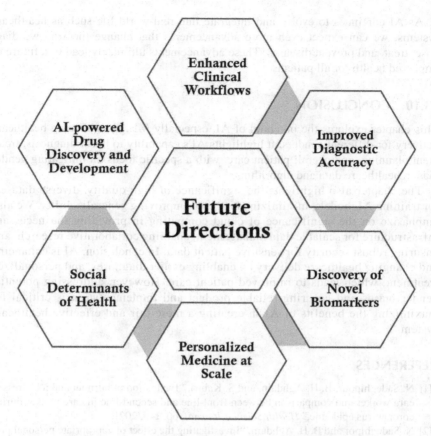

FIGURE 13.3 Emerging trends and future directions in AI-driven healthcare.

will continuously learn and improve, leading to earlier diagnoses and more personalized treatment plans.

- Discovery of New Biomarkers: Machine learning can play a key role in identifying new biomarkers associated with various diseases. This improves new diagnostic tools and targeted therapies that consider individual biological differences in disease presentation and progression.
- Personalized Medicine: Analyzing individual patient data, AI can tailor treatment plans by considering their unique biological makeup and predicted risk factors. This approach personalizes medicine, allowing for effective interventions.
- Drug Discovery and Development: Machine learning algorithms are changing drug discovery methods by analyzing big healthcare datasets such as molecular structures, patient data, and clinical results. This approach has the potential to have more effective and targeted treatments, tailored to the specific needs of different patient.

As AI continues to evolve and integrate into real-world life such as healthcare systems, we can expect even more advancements that change the way we diagnose, treat, and prevent diseases. These advancements ultimately lead to a future of improved health for all patients.

13.10 CONCLUSION

This chapter explores the potential of AI, especially ML, in enhancing healthcare delivery for different genders. It highlights AI's capability to boost diagnosis, treatment planning, and overall patient care, with a specific focus on mitigating gender bias in healthcare data and algorithms.

The chapter also highlights the significance of high-quality, diverse datasets for training AI models, minimizing bias, and improving generalizability. We also emphasize on the significance of cloud computing in providing the necessary infrastructure for scalable data management, enabling collaborative research, and ensuring robust security for sensitive patient data. In conclusion, AI is enhancing and changing healthcare delivery by enabling earlier diagnoses, and personalized treatment, which leads to improved patient care. However, addressing potential gender biases and ensuring reliable product and implementation is critical for maximizing the benefits of AI in creating a more fair and effective healthcare system.

REFERENCES

[1] N. Sadeghipour, B. H. Aghdam, and S. Kabiri, "Evaluation of burnout and job stress in care worker and comparison between front-line and second-line in care worker during coronavirus epidemic," *Health Science Journal*, pp. 1–5, 2021.

[2] N. Sadeghipor and B. H. Aghdam, "Investigating the effect of appropriate personal protective equipment on the stress level of care workers in the covid19 epidemic," *Health Science Journal*, pp. 1–5, 2021.

[3] E. Sadeghpour and E. K. Sangchini, "Assessment and comparative study of job stress in jam hospital jobs, tehran city," *Health Science Journal*, pp. 1–2, 2020.

[4] B. Mesko, *The role of artificial intelligence in precision medicine*, 2017.

[5] G. Chartrand, P. M. Cheng, E. Vorontsov, et al., "Deep learning: A primer for radiologists," *Radiographics*, vol. 37, no. 7, pp. 2113–2131, 2017.

[6] M. Borhani, M. R. Kalbassi, M. Psenicka, B. Falahatkar, and A. Crawford, "Identification and isolation of sprmatogonial stem cells from beluga (huso huso) testicular tissue," *Journal of Aquatic Physiology and Biotechnology*, vol. 10, no. 2, pp. 1–24, 2022.

[7] M. Borhani, N. Soofiani, E. Ebrahimi, and S. Asadollah, "First report on growth and reproduction of turcinoemacheilus bahaii (esmaeili, sayyadzadeh, ˝ozulug, geiger and freyhof, 2014)," *Zayandeh roud river, Iran. Austin Environmental Science*, vol. 2, no. 1, p. 1014, 2017.

[8] S. M. Sajjadi Mohammadabadi, "Advancing communication efficiency in electric vehicle systems: A survey of generative ai and distributed machine learning strategies," *Available at SSRN 4791891*, 2024, 23.

[9] S. M. Sajjadi Mohammadabadi, "Generative artificial intelligence for distributed learning to enhance smart grid communication," *Available at SSRN 4791903*, 2024.

[10] S. Avakh Darestani, A. Moradi Tadi, S. Taheri, and M. Raeiszadeh, "Development of fuzzy u control chart for monitoring defects," *International Journal of Quality & Reliability Management*, vol. 31, no. 7, pp. 811–821, 2014.

[11] S. M. S. Mohammadabadi, Y. Liu, A. Canafe, and L. Yang, "Towards distributed learning of pmu data: A federated learning based event classification approach," in *2023 IEEE Power & Energy Society General Meeting (PESGM)*, IEEE, 2023, pp. 1–5.

[12] S. M. S. Mohammadabadi, S. Zawad, F. Yan, and L. Yang, "Speed up federated learning in heterogeneous environment: A dynamic tiering approach," *arXiv preprint arXiv:2312.05642*, 2023.

[13] K. M. Kingsmore, C. E. Puglisi, A. C. Grammer, and P. E. Lipsky, "An introduction to machine learning and analysis of its use in rheumatic diseases," *Nature Reviews Rheumatology*, vol. 17, no. 12, pp. 710–730, 2021.

[14] N. Mehta, A. Pandit, and S. Shukla, "Transforming healthcare with big data analytics and artificial intelligence: A systematic mapping study," *Journal of Biomedical Informatics*, vol. 100, p. 103, 311, 2019.

[15] T. Yang, H. Zhu, X. Gao, Y. Zhang, Y. Hui, and F. Wang, "Grading of metacarpophalangeal rheumatoid arthritis on ultrasound images using machine learning algorithms," *IEEE Access*, vol. 8, pp. 67, 137–167, 146, 2020.

[16] S. Gurumoorthy, B. N. K. Rao, X.-Z. Gao, S. Shanmugam, and J. Preethi, "Design of rheumatoid arthritis predictor model using machine learning algorithms," *Cognitive Science and Artificial Intelligence: Advances and Applications*, pp. 67–77, 2018.

[17] S. Ho, I. Elamvazuthi, and C. Lu, "Classification of rheumatoid arthritis using machine learning algorithms," in *2018 IEEE 4th International Symposium in Robotics and Manufacturing Automation (ROMA)*, IEEE, 2018, pp. 1–6. *BIBLIOGRAPHY*, 25.

[18] J. Xie, Y. Li, N. Wang, L. Xin, Y. Fang, and J. Liu, "Feature selection and syndrome classification for rheumatoid arthritis patients with traditional Chinese medicine treatment," *European Journal of Integrative Medicine*, vol. 34, p. 101, 059, 2020.

[19] V. Venerito, O. Angelini, M. Fornaro, F. Cacciapaglia, G. Lopalco, and F. Iannone, "A machine learning approach for predicting sustained remission in rheumatoid arthritis patients on biologic agents," *JCR: Journal of Clinical Rheumatology*, vol. 28, no. 2, pp. e334–e339, 2022.

[20] D. Aletaha, "Precision medicine and management of rheumatoid arthritis," *Journal of Autoimmunity*, vol. 110, p. 102, 405, 2020.

[21] W. Tao, A. N. Concepcion, M. Vianen, et al.,"Multiomics and machine learning accurately predict clinical response to adalimumab and etanercept therapy in patients with rheumatoid arthritis," *Arthritis & Rheumatology*, vol. 73, no. 2, pp. 212–222, 2021.

[22] D. Castro Corredor and L. Á. Calvo Pascual, "Imbalanced machine learning classification models for removal biosimilar drugs and increased activity in patients with rheumatic diseases," *Plos One*, vol. 18, no. 11, p. e0291891, 2023.

[23] S. Momtazmanesh, A. Nowroozi, and N. Rezaei, "Artificial intelligence in rheumatoid arthritis: Current status and future perspectives: A state-of-the-art review," *Rheumatology and Therapy*, vol. 9, no. 5, pp. 1249–1304, 2022.

[24] N. Hammam, A. Bakhiet, E. A. El-Latif, et al., "Development of machine learning models for detection of vision threatening Beh͵cet's Disease (BD) using Egyptian College of Rheumatology (ECR)—BD cohort," *BMC Medical Informatics and Decision Making*, vol. 23, no. 1, p. 37, 2023.

[25] M. Maciejewski, C. Sands, N. Nair, et al., "Prediction of response of methotrexate in patients with rheumatoid arthritis using serum lipidomics," *Scientific Reports*, vol. 11, no. 1, p. 7266, 2021.

[26] T. Wei, B. Yang, H. Liu, F. Xin, and L. Fu, "Development and validation of a nomogram to predict coronary heart disease in patients with rheumatoid arthritis in northern china," *Aging (Albany NY)*, vol. 12, no. 4, p. 3190, 2020.

[27] M. Smuck, C. A. Odonkor, J. K. Wilt, N. Schmidt, and M. A. Swiernik, "The emerging clinical role of wearables: Factors for successful implementation in healthcare," *NPJ Digital Medicine*, vol. 4, no. 1, p. 45, 2021.

[28] J. A. DiMasi, H. G. Grabowski, and R. W. Hansen, "Innovation in the pharmaceutical industry: New estimates of r&d costs," *Journal of Health Economics*, vol. 47, pp. 20–33, 2016.

[29] K. B. Johnson, W.-Q. Wei, D. Weeraratne, et al., "Precision medicine, ai, and the future of personalized health care," *Clinical and Translational Science*, vol. 14, no. 1, pp. 86–93, 2021.

[30] E. Fosch-Villaronga, H. Drukarch, P. Khanna, T. Verhoef, and B. Custers, "Accounting for diversity in ai for medicine," *Computer Law & Security Review*, vol. 47, p. 105, 735, 2022.

[31] D. Cirillo, S. Catuara-Solarz, C. Morey, et al., "Sex and gender differences and biases in artificial intelligence for biomedicine and health-care," *NPJ Digital Medicine*, vol. 3, no. 1, pp. 1–11, 2020.

[32] Z. O. Merhi and L. Pal, "Gender 'tailored' conceptions: Should the option of embryo gender selection be available to infertile couples undergoing assisted reproductive technology?" *Journal of Medical Ethics*, vol. 34, no. 8, pp. 590–593, 2008.

[33] M. Mirbabaie, S. Stieglitz, and N. R. Frick, "Artificial intelligence in disease diagnostics: A critical review and classification on the current state of research guiding future direction," *Health and Technology*, vol. 11, no. 4, pp. 693–731, 2021.

14 Future Trends and Ethical Challenges in Transforming Gender Healthcare Using AI and ML

Rohit Khankhoje

14.1 INTRODUCTION

The ongoing discussion has clearly illustrated the emergence of a revolution in gender-based healthcare with, through, and around artificial intelligence and machine learning platforms. Primarily, the revolution is heavily anchored in addressing the unique gender-specific health needs and challenges [1].

With AI and ML heavily exploiting the massive datasets about gender-specific health conditions, the patterns, correlations, and risk factors in some diseases have become uncovered. In diagnostics, AI enables more accurate screening for certain diseases, such as mammograms hence early detection and improved patient outcomes. For treatment, the approach is growing more personalized and taking advantage of the most astounding gender-specific genetic differences. AI gives practice the leverage of optimizing treatment to be more effective and less sidelining as much as possible. All of these applications are a total turn-around from the conventional one-size-fits-all model, which is just great! In prevention care, AI applications give patients the authority to take control of their health in the most proactive and awe-inspiring way. For example, through machine learning, a wearable device, or a health app one may monitor his or her blood sugar and give fitness, dieting, and other equally outstanding recommendations.

However, there is a need to be fully aware of the trends and ethical challenges coming ahead. It is only with this knowledge that one can best utilize these amazing potentials of AI and ML in gender-based healthcare in a most remarkable way. This chapter, therefore, is a genuine effort to the readers to give an exhaustive understanding of the upcoming area in healthcare with AI and ML. It gives insight into the most promising trends and ethical choices to be made in the future. Therefore, the chapter emphasizes the importance of maintaining a commitment to a balanced approach that fully exploits the revolutionary aspects of technology while preserving ethical concerns.

DOI: 10.1201/9781003473435-14

14.2   EMERGING TRENDS IN GENDER-BASED HEALTHCARE

The healthcare domain is currently undergoing a notable transformation driven by groundbreaking technological innovations that provide the opportunity to improve patient health, streamline operational processes, and enhance the overall delivery of healthcare services. This chapter delves into the latest technological trends shaping the future of healthcare as shown in Figure 14.1.

14.2.1   TELEMEDICINE AND REMOTE PATIENT MONITORING

Telemedicine is a new method of providing medical services from a distance using telecommunications technologies. Virtual appointments are the basic requirement, which means that the patients' side entails video calls, phone calls, or messages.

Then, it implies medical advice, diagnostics, and prescriptions being given without face-to-face visits [2]. Additionally, telemedicine encompasses remote monitoring, allowing patients' essential signs and conditions to be tracked at all times. Patients' engagement is promoted with mobile health that includes self-management, medication reminders, as well as the opportunity to be linked to learning resources [5].

On the other hand, the impact of telemedicine is tremendous. It eliminates the geographical boundaries and enables access to distant and remote areas. By facilitating virtual appointments, it alleviates patients' need to wait in the traffic or come and depart at the premises [3]. At the same time, it also allows patients to save up on the costs of travel, sleep, and time off work. Moreover, it was invaluable in the emergency set as it allowed care delivery when people were to avoid direct contact at all costs [6].

Remote patient monitoring implies collecting data outside traditional care settings. It is delivered using secure gear, such as wearable monitors, transmitting patients' data in real-time. In case of abnormalities, notifications warn clients in advance and

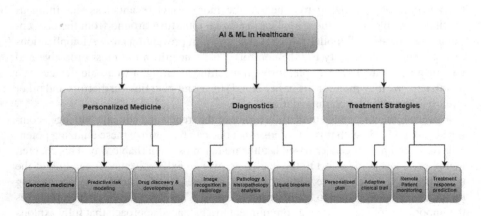

FIGURE 14.1   Emerging trends in healthcare.

offer alerts. RPM's effect is transformative in the sense of empowering early detection. An estimated 70% of all hospital consultations could have been avoided if the routine monitoring of chronic patients was more regular as offered by remote means. It aims to enhance patient quality of life provided in a patient-friendly package [4]. Both telemedicine and remote patient monitoring revolutionize the way care is delivered. They make it more accessible, more convenient, and more patient oriented. They not only offer a successful alternative but also enhance the development of preventive medicine and chronic care health. This mix ensures that care delivers timely, customized, and available services.

14.2.2 INTERNET OF THINGS IN HEALTHCARE

IoT devices are widely employed to track patient health and control chronic disorders and enhance preventative care. Wearable units, intelligent sensors, and interconnected medical instruments gather actual-time data that empowers medical individuals to make decisions and lets patients be more engaged in their healthcare. Utilization of the Internet of Things (IoT) in the health sector due to the prospect of a sizeable greater number of units connected creates a network of interconnected devices to gather, deliver, and analyze data to enhance health outcomes. This novel approach exploits digital technology to remodel traditional healthcare delivery and deliver a more responsive and patient-centered application [7].

14.2.2.1 The Role of IoT in Healthcare

Remote Monitoring: IoT devices like wearable heart monitors, glucose monitors, and even fitness trackers allow patients to be monitored even beyond clinical settings. The simultaneous data collection allows one to manage one's health reactively, particularly with chronic conditions such as diabetes and heart disease, where a patient can be forced to seek medical services when it is too late.

Augmented Patient Engagement and Adherence: IoT devices give the patient more control over their health in the process. Intelligent medication dispensers and wearable technologies act as reminders to the patient to take their medication as prescribed, as well as the amount of exercises to take, which would consequently increase one's compliance to the treatment procedures to a larger percentage [8].

Data-Driven Insights for Personalized Care: IoT devices generate vast amounts of health data. When analyzed using artificial intelligence and machine learning, such data can provide deep insights into patient health patterns, allowing for personalized and proactive healthcare [9].

Operational Efficiency: In the case of hospitals, IoT technologies such as smart beds, connected inhalers, or RFID-marked tools help to simplify processes, optimize asset application, and reduce costs. This way, such technologies improve care for patients and increase the efficiency of medical facilities at the same time.

14.2.2.2 New and Coming in IoT for Healthcare

The fifth generation technology has the ability to bring a drastic change in the IoT realm especially in the healthcare industry. It can lead to a faster, more reliable and more efficient data transfer, improving the operations of telemedicine and remote monitoring [10]. Therefore, it can lead to more practical video appointments in real time and remote operation. The future of hospital services sees the light in smart hospitals that have full integration of the Internet of Things. These hospitals take advantage of this technology to monitor patients, regulate inventory, monitor the environment and more. This implies that these hospitals offer the most efficient services and the highest level of patient care. The application of IoT is extended to mental health services through apps and wearable devices. The devices track several parameters such as sleep patterns, physical movements and variations in heart leading, which the specialist uses to draw insight into the mental status [11].

The promotion of Internet of Things use has led to increased concerns on data hacking and privacy invasion. Therefore, sophisticated security measures and layers of encryptions were advanced to protect this sensitive health data. By retrospectively evaluating the vast amounts of data from the use of IoT devices, the clinicians can find early signs of disease exposure and early signs of exposure. This implies they can be guided to enroll to avoid exposure or institute a lifestyle change. In brief, the Internet of Things in medical applications promotes not only technological advancement but a more efficient, personal and achievable health processing system. As we look into the future, the principle of Internet of Things technologies with potential can improve the caliber of healthcare delivery and outlook.

14.2.3 VIRTUAL AND AUGMENTED REALITY IN MEDICAL TRAINING

Medical training and education have been revolutionized with the introduction of virtual and augmented reality technologies. It allows specialists to virtually engage in practical training, creating simulations that help them improve their skills and initially qualified practical experience in a risk-free setting. This incredible technology has revolutionized surgical training and many other challenging medical conditions. Virtual reality (VR) is a driving force in the medical revolution. The use of VR technology to train and improve the abilities of healthcare personnel is indeed groundbreaking. VR technology can build interactive 3D worlds of full immersion in which medical trainees may experience gameplay in various clinical scenarios. Trainees might perform nearly anything, from a physical evaluation to complicated surgery. This command of the gameplay is critical to accomplish their skilling and make their procedural confidence without posing a risk to the patient. Because the technology allows them to try the process again, trainees might develop more difficult surgical treatments or diagnostic operations and carry them out in a secure and constrained environment. In addition, adaptable learning tools can be used, allowing teachers to administer various clinical scenarios to fit particular training demands [12].

This elasticity adds depth and breadth to the training, preparing the trainees for a much more varied range of real-life health problems. Furthermore, this technology is highly inexpensive and has low resource demands and is risk neutral. For this reason,

VR medical training isn't only a groundbreaking instructional method; it is also a fundamental redefinition of how to master, teach, and prepare for the complicated patient treatment.

Medical training, on the other hand, is being reoriented by Augmented Reality, which modifies the learning and skill-acquisition process for medical personnel. Its combination with digital information superimposed on the real world stimulates auditory and visual learning by providing complex 3D visualization, thus changing the way anatomy education and even surgical training take place [13]. For example, complex anatomy and surgical concepts can be projected onto a mannequin or patient's body to provide an experience-rich, realistic image. Instead of vague visualization of the human body structures, this technology gives students and practitioners an opportunity to materialize bodily systems and surgical scenarios. Therefore, they have a clear link between an academic understanding of the concept and implementing it. For surgery, augmented reality (AR) aids in providing visual direction and guiding directions, ensuring accuracy or precision and reducing errors. It integrates visual or auditory feedback in real-time, ensuring the intended course of action. AR in medical education not only enhances knowledge of human anatomy and an array of medical processes but also gives learners the ability to have an experience-rich process for learning. This type of learning design is beneficial to daily procedures and rare surgical procedures because it ensures that a specialist, despite the experience or stature, has undergone thorough training. Consequently, AR is propelling the medical teaching sector forward. It further contributes to better patient outcomes by ensuring educated and trained personnel. VR and AR have advanced medical education. While they improve the learning process, they have a resultant effect on the patient's welfare and the cost of training. They also guarantee the continued growth of the professionals in the diverse field of medicine.

14.2.4 Robotic-Assisted Surgery

Robotic-assisted surgery is defined as surgeries in which surgeons use robotic systems to perform surgical tasks with a higher level of accuracy and command than human ability. A "surgical system" generally includes a console, a patient-side cart equipped with robotic arms, and a high-definition 3D vision system [14]. Robotic arms are controlled by the surgeon, who directs them as they transpose the surgeon's natural arm and wrist movements into finer and more precise movements of tiny instruments inside the patient's body [15].

14.2.4.1 Emerging Advancements in Robotic-Assisted Surgery

Amplified Precision and Flexibility: Contemporary robotic systems provide amplified precision and a wider range of motion compared to human hands, hence allowing for more complex and less invasive procedures.

Three-Dimensional Imaging and Augmented Reality: sophisticated imaging, producing live, three-dimensional, on-screen views of the surgical area. In addition, augmented reality overlays can be superimposed to provide appropriate additional data, for example the position of blood vessels applies to the use of Flinders Technology Associates—robotic assisted surgery device.

Integration of Machine Learning and Artificial Intelligence: incorporate machine learning and AI to help surgeons. For instance, the latest da Vinci robot incorporates AI and machine learning to analyze and recommend surgeries using a vast database of various forms of such systems that have been used in thousands of procedures. This has the potential to cut down the time spent in surgery and improve outcomes.

Telesurgery Capabilities: Recent advancements encompass the ability to perform remote surgeries, hence making specialty surgery more accessible applies to the telesurgery capability of the world's only advanced laparoscopic robotic systems sub-specialised for colectomy [16].

Haptic Feedback: Ongoing developments include systems that provide haptic feedback, so the surgeon can feel what they are performing.

14.2.4.2 Implications for Healthcare

First, outcomes for patients are enhanced. Because robotic-assisted surgeries are generally less invasive, patients tend to experience less postoperative pain, blood loss, scarring, and have shorter recovery times. This increased survival rate can change the patient's quality of life and mitigate the significant illness burden. Second, it boosts the surgeon's performance: the devices are ergonomically developed, lowering surgeon fatigue and minimizing the danger of surgeon errors during lengthy or difficult operations. Third, the operation time and length of stay are reduced, leading to shorter hospitalization and reduced surgery time and shorter overall hospitalization [17].

Training, education, and instrument challenges: the robotic device can duplicate surgical operations for training and be utilized as an educational tool by surgeons. Nevertheless, robotic-assisted operations are extremely expensive, need an extended learning curve for the surgeon, and the majority of the instrument must be recurrently maintained and updated. Automated operations are a significant milestone in surgical procedures and offer excellent benefits to patients and the healthcare workers. Automations in healthcare have and shall continue to grow with the progressive improvements in robotics, machine learning, and imaging. Given the significant benefit of safe, minimally invasive, and less surgical interventions.

14.2.5 CHATBOTS AND VIRTUAL HEALTH ASSISTANTS

Powered by incredible advancements in artificial intelligence algorithms, chatbots and virtual medical assistants can offer instant responses to patients' questions, medication reminders, and help monitor chronic diseases. These innovative technologies substantially increase patient involvement, provide continuous help, and are crucial to health [18]. A chatbot and virtual health assistant in the health sector is an interaction system driven by artificial intelligence created to perform human-to-human interactions autonomously. This tool relies on theoretical concepts such as natural learning processing and machine learning to understand and interpret spoken human languages and behave in a way that seems more human than if it was operated by a simulation of prearranged instructions [31].

14.2.5.1 New Advancements in Chatbots and Virtual Health Assistants Include

Enhanced Natural Language Processing: Modern chatbots are embedded with advanced NLP abilities that help them understand and respond to complex medical questions more accurately.

Personalization and Machine Learning: Chatbots are capable of learning from conversations, making their responses more accurate and relevant the longer they are used.

Integration with Electronic Health Records: Some chatbots are synced with Electronic Health Record systems, providing them with access to patient records, filling down appointments, and recommending medical care tailored to individual patient needs.

Multifunctional Devices: In addition to answering questions, chatbots are also able to conduct symptom checks, notify patients when they need medication, offer mental health support, and provide post-operative care instructions.

Voice Interactions: Most modern virtual assistants are equipped with voice-activated technology for a more natural and easier mode of communication.

Telehealth applications: Virtual assistants are being used in telehealth to make virtual visits and triaging patients more efficient.

14.2.5.2 The Impact of Chatbots and Virtual Health Assistants on Healthcare Includes

Enhanced patient engagement and support: Chatbots operate around the clock, responding to questions and providing health information. Thus, patients have improved support and engagement on a 24/7 basis.

Improved healthcare access: these systems improve access to healthcare, especially when minor inquiries are involved, or in specific regions where health accessibility is a challenge.

Healthcare delivery efficiency: chatbots automate interactions and tasks that would be done manually by medical staff. Subsequently, workers are freed from routine tasks, allowing them to focus on more critical responsibilities, creating an efficient system.

Data gathering and analysis: chatbots help collect information from patients which can be analyzed to comprehend patient needs and health trends.

Cost-saving: considering chatbots automate interactions and offer preliminary support, they result in cost savings for providers.

The challenges of chatbots include ensuring medical information is accurate. This is due to privacy and patient confidentiality concerns, such as cases of breached data, and the impact of reduced human interaction [19].

14.3 APPLICATIONS OF AI AND MACHINE LEARNING IN HEALTHCARE

14.3.1 PERSONALIZED MEDICINE

14.3.1.1 Genomic Medicine

Genomic healthcare represents a radical shift in the provision of healthcare by using Artificial Intelligence and Machine Learning to investigate the intricate features of human genetics. The reshaping factor is AI and ML's potential to scrutinize substantial genomic datasets professionally, a futile effort with traditional methods, primarily due to the complexity and scale. They shine in unveiling complex patterns and variances in individual genomes, both of which are frequently vital for realizing one's predisposition to disease and reaction to various treatments. Genomic medicine with AI and ML is paving the way for ultra-customized healthcare. Synergistic variations allow distinct treatments to address the precise needs of one's genetic composition. This degree of immunity also means that treatments and drugs may be modified to the nuances of each patient's genetic makeup, drastically enhancing their performance. For example, genomic healthcare in the cancer sector can help identify precise genetic alterations that are responsible for the disease, and the appropriate treatment is more probable to be effective and less likely to create adverse effects [20]. A rather intriguing, if not astonishing, utilization of the technology that

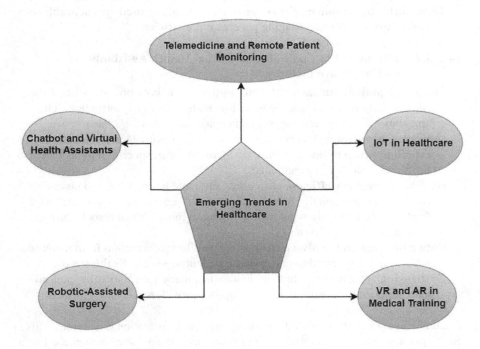

FIGURE 14.2 Application of AI/ML in healthcare.

extends far beyond treatment as well as into prevention is being turned. By learning genetic features for predetermined risks, AI can predict the risk of contracting specific diseases if warnings are not detected early. The medical sector, which has been rooted with the usual system for centuries, must make a big adaptation.

14.3.1.2   Predictive Risk Modeling

Predictive risk modeling as enabled by AI algorithms is a considerable departure in the landscape from reactive healthcare to preventive healthcare. It involves a deep analysis of an individual's genetic and clinical data to predict the likelihood of certain diseases. AI algorithms are the best pick for the job because they have the capacity of analyzing vast and complex data sets. The algorithms can be fed with large health records, genetic information, lifestyle factors, and environmental data, among others, which they analyze and establish patterns and correlations that suggest future health risks [21].

The beauty of prognostic risk modeling is its ability to identify individuals at an advanced risk of getting certain diseases. This identification is critical as physicians can put in place proactive measures to prevent or contain the disease. For example, an individual identified as at a high risk of getting heart disease would be advised to live well, be doing checkups, and take a certain medicine just to be safe. Perhaps the most critical aspect of prognostic risk modeling is that it cuts across physical health to mental health, which allows for early interventions that could make the difference in managing such conditions. Predictive risk modeling improves the quality of healthcare by offering personalized risk assessment services. This move from reactive to proactive care translates into better health outcomes while it can reduce the ultimate cost of healthcare by avoiding or mitigating severe health events.

14.3.1.3   Drug Discovery and Development

In the sphere of pharmaceutical exploration and development, Artificial Intelligence is an essential change driver, substantially accelerating and improving the process. Specifically, AI's scope is in its capability to instantly model the complicated interactions between various molecules, which assists researchers in recognizing auspicious drug candidates from an extensive pool of options. When equipped with machine learning algorithms, AI can predict not just how distinct molecules will interact but also the way great candidate molecules are at curing agents [22].

This predictive ability is especially beneficial in developing drug possibilities, which is critical in managing an extensive spectrum of illnesses. One of the major benefits of artificial intelligence to the pharmaceutical manufacturing procedure is a substantial reduction in the time it takes to create and the cost of inventing drugs. The act of producing a novel drug has historically lasted over a decade and has been quite costly to bring to the market.

However, data analysis allows AI to perform the initial phases rapidly, eliminating the need to conduct some early trials that may be quite expensive. Another reaction that AI has vigorously impacted is personalized medicine. AI uses patient data to design medications that are modified to a patient's genetic profile, boosting their usefulness and decreasing the possibility of side effects from therapy. Overall, synthetic argumentation in the field of drug discovery and enhancement will not only hasten the time it takes to bring fresh remedies to market but also increase the progressions' efficiency and usefulness. It has transformative potential for public health and the capacity to address some of the most severe medical problems.

14.3.2 DIAGNOSTICS

14.3.2.1 Image Recognition in Radiology

With the advent of machine learning algorithms designed for image recognition, radiology has made tremendous progress. Artificial intelligence, applied to various medical imaging such as X-rays, MRIs, or CT scans, shows fantastic results in its analysis. Moreover, the opportunities that AI opens in radiology exceed those of a human pathologist. AI analyzes complex visual information, quickly identifies various abnormalities, tumors, and other pathologies at an initial stage, which cannot be noticed or distinguished by a human being: "Many studies have shown that compared with human radiologists, these algorithms can detect these findings in imaging studies more accurately and reliably" [23].

AI effectiveness in image recognition is not limited to the detection of pathologies. It can also localize and measure these abnormalities and in some cases even allow their initial differentiation. The potential opportunities in oncology are immense. If AI tools allow for early, quick, and accurate identification of tumors, the effectiveness of cancer treatment will increase substantially. Moreover, AI facilitates the work and significantly accelerates the diagnostic process. If before, some studies took an hour or more for a radiologist, with the use of AI, the algorithm can process this info in a matter of seconds. The acceleration of the diagnostic process can save the life of the patient in emergency medicine and the right decision-making policy regarding his disease. Although AI will be a powerful tool for improving diagnostic and clinical therapy services, it will not replace professional-trained radiologists. Therefore, the use of AI in radiological imaging is a new revolution in contemporary medical science.

14.3.2.2 Pathology and Histopathology Analysis

Artificial intelligence has become a valuable tool in the field of pathology and histopathology and has significantly increased pathologists' capabilities. Artificial intelligence systems that are specially designed for repetitive tasks depending on advanced machine-learning algorithms can analyze histopathological images in excellent detail and recognize patterns or variations that might imply a variety of diseases but especially cancer [24].

The following algorithms have been trained on extensive, annotated image databases which considerably differentiate small changes in the provided tissue samples that may go unnoticed by human observation. The impact of this technology is deep, particularly when it comes to early detection. For example, when it comes to cancer diagnosis, AI can detect tiny malignancies or precancerous changes in tissue slides, often catching such conditions at an earlier stage than traditional methods do. Early detection is critical because it improves outcomes for patients by allowing expedited and targeted therapies.

Additionally, AI significantly contributes to the diagnosis's precision. It gives quantitative measurements for each histopathological criterion, eliminating the subjectivity factor associated with manual microscopic examination. This kind of accuracy is extremely beneficial for complicated cases where even minimal histopathological information is important in determining the correct diagnosis and therapy planning. Additionally, AI processes a large number of samples very quickly; this can reduce the tremendous burden on human pathologists so they can focus on the more complex cases, decreasing the turnaround time for test results. Thus, AI does not replace the knowledge or expertise of human pathologists but instead supplements it. By enhancing the accuracy and efficiency of pathological examination, the overall level of patient care available to smaller care facilities is substantially enhanced.

14.3.2.3 Liquid Biopsies

Liquid biopsies are a disruptive innovation in cancer diagnostics in that the interpretation of complex biological data is performed by artificial intelligence. Liquid biopsies focus on the analysis of circulating biomarkers, which may be tumor DNA released by the tumor into the bloodstream in the form of fragments or cells. Unlike typical biopsies, which are highly invasive, this makes for a perfectly non-invasive option, and it is much more comfortable for the patient. Artificial intelligence in this case is the possibility to analyze vast amounts of complex data from these various biomarkers. Artificial intelligence algorithms trained on large datasets can detect subtle patterns that arise in genetic mutations associated with various cancers. The importance of such a view is that it allows for the early detection of cancer, often before any symptoms present, or before it spreads further and shifts to other areas of the body. Moreover, liquid biopsies combined with AI analysis allowed for watching over the evolution of cancer and its reaction to treatment. In terms of custom-tailored treatment planning and minor adjustments based on important feedback, this becomes a perfect trend. In general, this option affects cancer treatment and monitoring the patient from being uncomfortable and invasive to perfectly non-invasive and patient centered. This effect is demonstrated by helping detect cancer early, develop personalized treatment plans, and improve patient outcomes.

14.3.3 Treatment Strategies

14.3.3.1 Personalized Treatment Plans

At the same time, the introduction of AI in the sphere ensured a considerable shift in the novel paradigm of operations: thus, the method of treating has transitioned from general ones to more personalized ones. AI technology allows creating such treatment tactics which are entirely novel or updated in reference to the patient's individual pattern and conduct, depending on his/her genetic composition and doctor's behavior [25].

The reason for this is that a plethora of patient data is analyzed: genetic profile, reaction to prior treatments, and consistent health monitoring data. Using machine-learning algorithms that can dissect and convert this often multi-faceted data, trends, and connections not otherwise evident through consistent research. That is how optimal success and quality ones stand based on the request for the patient and genetic structure. The effect of such individual treatment approaches is evident—on the one hand, patients obtain a therapy that would be the most effective in their individual case, thus increasing efficiency. Moreover, such approaches aid in reducing the number of side effects often accompanied by more unspecific treatments. Therefore, AI-assisted treatment plans aid in the current expansion of the latest direction of the contemporary model of healthcare.

14.3.3.2 Adaptive Clinical Trials

Adaptive clinical trials are a significantly innovative approach to the field of medical research, employing machine learning algorithms to develop a trial design on the move. It is a shift away from conventional fixed-trials in design and enables greater flexibility and, therefore, efficiency. Adaptive trials use the algorithms whereby the real-time data from the trial participants collected are analyzed continuously. The data stream permits us to adjust the trial in progress without compromising its scientific validity.

Critical adjustments in adaptive trials included the optimization of doses, patients can be reassigned within a treatment group, or one or more treatment eyes could be added or removed based on the knowledge generated up to that point on the fly. In other words, this feature relies on real-time data analysis to identify effective treatment quickly, eliminating the extensive time and resources pilfered by the normal process. Also, it facilitates the reduction of patient exposure to more hazardous or less effective interventions [26]. Secondly, adaptive clinical trials are inclusive of a broader base of study population. Taking into consideration the diverse range of clinical trials population, we can question whether the specific population is the same as the general population. Inclusion considerations assist in generalization for universal applicability.

In conclusion, the use of machine learning algorithms in adaptive clinical trials best exemplifies a resilient reactive and patient-based research approach. It is critical in accelerating the pace of medical breakthroughs to develop safer and more effective treatments. Ultimately, better treatment plays a purpose in creating a healthier community.

14.3.3.3 Remote Patient Monitoring

Remote Patient Monitoring implies a significant leap towards the development of healthcare, in terms of efficacy and complexity that artificial intelligence brings. The method consists of using AI-controlled devices to monitor patients beyond the hospital or clinic walls on a regular basis. They include all kinds of devices from portable sensors to daily use gadgets for monitoring, gathering loads of health data such as vital signs like pulse, blood pressure, drug adherence, and even life-related data such as physical activity levels [27].

This information is analyzed in real time with the help of AI algorithms, providing the health-caring staff with the opportunity to watch the patients' current health state thoroughly. The data flow also allows detecting possible health problems early, before they turn into something serious. For instance, AI may notice slight vital sign changes that may signal a complication starting to develop, allowing intervening immediately. Personalization of care also is implemented with the help of RPMs. Monitoring patients' individual responses to treatments helps change the therapy plan in the blink of an eye, making sure that it is tailored to each patient's unique health needs. This ensures the therapy is effective, but it also makes patients more involved in the process and they're more likely to follow the therapy plans. In summary, RPM combined with AI is changing the way healthcare is provided. It eliminates the border between patients and the system creating a patient-oriented proactive care that is built on a consistent nurturing relationship that significantly improves the patient's health and quality of life.

14.3.3.4 Treatment Response Prediction

The implementation of machine learning models has transformed the panorama of personalized medicine by the introduction of treatment response prediction. This application employs past health data to predict how patients may respond to certain treatments. By tracking and analyzing previously recorded patients' data, encompassing demographic characteristics, genetic factors, and medical history, as well as their outcome following treatment, ML algorithms can forecast and approximate the efficacy of a given therapy to new patients. This capability has a phenomenal impact on the health profession regarding the identification of optimal treatment.

Models for prediction of treatment response are advantageous to healthcare providers due to two primary reasons. Firstly, the conventional trial-and-error strategy was not only inefficient but also exposed patients to potential harm. The new approach is guided and focused, significantly eliminating the disparity between the treatment duration and effective response time. A major application of ML in this context is in cancer analysis. Such applications can estimate how a patient's cancer will react to diverse chemotherapy drugs, thus enabling oncologists to use the drug associated with maximum odds of success [28].

Secondly, this method significantly reduces the risks of adverse reactions and associated side effects. By selecting drugs that are most likely to be compatible with the patient's unique body, the rate of adverse drug reactions is greatly minimized. Ultimately, personalized medicine accompanied by prediction models has transitioned the treatment approach from universal to the unique level.

14.3.4 Anticipated Future Developments

The future of gender-based healthcare is just on the horizon of revolutionary changes driven by the incorporation of Artificial Intelligence and Machine Learning. Predictive analytics, genomics, and patient-specificity aspects will merge into a novel quality of healthcare that is based on the specifics of a patient's gender and promise an optimistic outlook on the future trends.

14.3.4.1 Predictive Analytics in Gender-Specific Health

Predictive analytics driven by AI will be vital in predicting and preventing gender-related health threats. Using machine learning algorithms, large volumes of data will be mined searching for distinctive patterns, predisposing factors, and early warnings for diseases that affect each gender disproportionately. Such a proactive approach will allow users to develop measures to counter those threats before they even emerge, which will ultimately lead to better outcomes [25].

14.3.4.2 Genomics and Precision Medicine

The future of gender-based health is on the verge of a breakthrough transformation via genomics and precision medicine. Advanced AI and ML capabilities allow for analysis of one's genetic profile with respect to gender disparities. This means that a treatment plan will be tailor-made for each individual taking into consideration one's gender. This approach will not only increase efficiency but work wonders in eliminating gender disparities in the field. Genomics will integrate seamlessly into the clinical practice process to become a foundation of precision medicine.

14.3.4.3 Patient-Centric Approaches

AI and ML will revolutionize the present and future state of healthcare with patient-centric approaches to prioritize specific individual needs and preferences. Advanced and novel predictive modeling methodologies will quickly analyze health risks and treatment preferences for males and females, respectively. With cutting-edge virtual health assistants powered by the ML algorithms, healthcare chatting will become much more secure and effective not to create and deliver unattainable health insights, gripping timely notifications and reminders, and much-needed pervasive backing. Such unwavering commitment to patient-centric approaches will undoubtedly boost patient engagement, improve treatment plan adherence, and yield the highest patient satisfaction with healthcare services.

14.3.4.4 Breakthroughs in Women's Health Research

The combination of AI and machine learning will confidently create a new era of breakthroughs in women's health research. Advanced technologies and algorithms will help us identify new treatments, diagnoses, and interventions a thousand times faster than those patients with cancer are diagnosed and given some palliative advice. From these vast sets of highly structured and unstructured information, we will uncover unidentified trends and patterns minutes later. As a result, one will almost immediately understand what a unique problem women face regarding their health.

14.3.4.5 Innovations in Remote Monitoring and Telehealth

The future is unfolding with remarkable improvements in remote monitoring and telehealth to match the specific health needs for each gender. Advanced wearable devices powered by AI will collect and analyze patient-specific information in real-time, delivering critical knowledge instantly useful for both patients and physicians. These proactive monitoring procedures will be timely warning most likely health concerns, conductible interventions and substantially improve healthcare accessibility at large.

In conclusion, the future directions in AI and machine learning applications for gender-based care depict a transformative era of precise, personalized, and patient-centered care. The combination of predictive analytics, genomics, and patient-centered care approaches has the power to revolutionize the landscape of care delivery by ensuring that all gender-specific health needs are comprehensively cared for. However, as these developments continue to unfold, ethical considerations and responsible implementation will be integral in realizing the full benefits of AI and ML in gender-based care.

14.4 ETHICAL CONSIDERATIONS IN FUTURE HEALTHCARE

The surge of new technologies raises multiple ethical concerns that mankind should address. In this discussion, we will concentrate on three critical areas of concern, data privacy, informed consent, and algorithmic biases.

14.4.1 DATA PRIVACY: STRIKING A BALANCE BETWEEN INNOVATION AND SAFEGUARDING

One of the central ethical questions of the prospering technological developments is related to data privacy. As AI and IoT become more integrated into daily life, they are increasingly dependent on massive amounts of people's information. The issue to fight deals with the opportunity to find the balance between a need to foster innovation and privacy protection. To offer more personalized services and frictionless experience, companies collect a massive amount of information about users. This information includes not only browsing habits but even health statistics that is a valuable resource for developing cutting-edge technologies. However, the ethical dilemma is that people remain unaware of how their data is being used or do not have adequate means to control the ways that information is shared [25].

Therefore, to address this specific issue, robust and reliable privacy policies, transparent data practices, and decisiveness from developers and technology vendors will be of utmost importance. Ethical frameworks should require that users have access to meaningful autonomy, ensuring that individuals understand how their personal data is used, govern it, and consent.

14.4.2 INFORMED CONSENT: NAVIGATING COMPLEX CHOICES

Informed consent is highly relevant to many spaces, especially as technology continues to evolve and diversify. Many technologies from predictive analytics in healthcare to decision-making machines operate beyond the control of the user. The ethical issue at play is the extent that users need to understand the full implications of their use of a given technology. Many technologies are complex and make it challenging for users to understand their work entirely.

Moreover, the technology itself may change faster than users can understand. To address this, developers and other stakeholders should leverage options for communication. Should documentation, guides, and user interfaces describe what the technology does in an understandable fashion? Moreover, do users have a way to study how the technology works with provided materials? Finally, regulators should keep up to date based on technological advancement.

14.4.3 ALGORITHMIC BIASES: UNRAVELING ETHICAL QUANDARIES

Artificial intelligence has brought the issue of algorithmic biases to the forefront. Machine learning algorithms train using living history data. If there are career biases in the data, the algorithms will train on them and then reinforce them in the decision-making process.

Algorithms can be biased in different domains such as hiring practices, loan approvals, law enforcement, and healthcare. As a result, marginalized groups may be more susceptible to developing inequities in the various systems. Failure to address biases in algorithms contributes to a digital divide and compromises fairness and justice systems.

To conclude, developers and policymakers should work together to minimize the risks associated with algorithmic bias. Developers have a responsibility to check training data for bias, apply methods for reduction and conduct comprehensive assessments of the fairness of the outputs. Additionally, ethical guidelines or standards to regulate the use of AI should also be developed to oversee what the AI should do and apply fairness, accountability, and transparency. In summary, the ethical questions associated with emerging technologies are complex and designing ahead should reflect several considerations. First, data privacy, consent and algorithmic fairness should be scrutinized when making such ethical decisions. We need to set up everyone for civil liberties, enabling a condition where consumers can keep control of their data and know the implications of this technology. Second, informed consent should also be dynamic and continuous monitoring, depending on new uses of technology or exposure to new risks. Lastly, to address algorithmic biases is not a technical problem alone but a social one. How can developers, policymakers and the society as a united unit ensure fairness and eliminate biased histories created using intelligent systems is a challenge.

14.4.4 BALANCING INNOVATION WITH ETHICAL RESPONSIBILITY

One of the most disputable issues remains the balance between the search for an innovative idea and the increased ethical responsibility.

In the current rapidly changing high technology environment, it is essential to keep in balance the aspiration to break the limits of what is possible and browse the ethical line that protects people's rights and guard society's values. It is also crucial, as innovation leads to progress, ensures more productivity, and sometimes results in social benefit, although violations, such as privacy, security, fairness, and moral considerations are also raised.

Therefore, success requires from the company to build a balance between the value of innovation and a value-based trend in social responsibility. The real source of this balance is the ethics that is integrated into the very core of the innovations. It is essential to evaluate the new product idea's impact on key stakeholders, consider society's consequences in the coming years, and remember issues such as data privacy and security in the rapidly evolving tech world.

14.4.5 STRATEGIES FOR ACHIEVING EQUILIBRIUM BETWEEN INNOVATION AND ETHICS

Varied and Comprehensive Data: The utilization of varied and comprehensive data is crucial for making AI models in gender-sensitive healthcare efficient and fair. First, it includes the collection of various kinds of data to represent every gender and gender expression as accurately as possible. Secondly, when such comprehensive sets of data are used for AI model training, the formulated algorithms are better trained to understand and address the diverse and unique healthcare needs associated with different gender experiences. Such diversity in data helps make AI-related healthcare solutions more precise and relevant, eliminates the danger of gender bias, and guarantees that the developed advanced tools provide fair and effective care to everyone.

Interdisciplinary Method: An interdisciplinary method for the development of AI systems for healthcare, mainly gender, would also be important in ensuring that the AI systems maintain ethics. This would mean diverse forms of experts, including, medical practitioners, ethicists, sociologists, and members of the queer community. By bringing the expertise together, the end product, in this case the AI systems, would be developed to be as sensitive to the complexity around gender as possible. Anything less than this approach would mean that such systems are not just medically accurate but also weak ethically and uninformed culturally, and are biased and only developed to treat only a section of the population [29].

Regular Evaluations for Bias: Scheduled checks for issues like bias for the AI system, especially those systems with gender parameters, is also a crucial ethical consideration. This would mean that the AI developers must have a regular schedule for monitoring and evaluation of the AI algorithms for bias based on gender. This would be important because bias would interfere with the outcomes of the system and influence unfair or unresourceful decisions. For as much as bias is inevitable, where it is detected earlier, it can be resolved. This means that the AI systems remain fair to all men and women and will be free from bias in all instances. This is critical in terms of AI principles in ethics, especially in issues to do with health and data privacy.

Robust Data Management: Data governance in the applications of artificial intelligence and notably in the healthcare sector would need some robust data and protection policies are anything goes into the patients' consent, privacy, and security, and responsible patient data use. It would also involve the protection of their privacy from unauthorized access such data breach and cybersecurity issues. Such practices of data governance would help in appearing ethical across all applications, and that would also help the patients feel some level of trust due to the privacy-centric issues which are a matter of confidentiality.

Ethical Supervision: Establishing a committee pertaining to ethical supervision is a critical measure related to the monitoring of AI and ML application, especially in the context of gender-oriented healthcare. This committee should consist of professionals specializing in ethics, technology, and medicine, as well as potentially representatives from different gender backgrounds to ensure personal experience. It will be responsible for overseeing development and usage of AI applications, as well as reviewing and providing guidance on AI projects in terms of suitable ethical considerations, such as bias, patient awareness and privacy, and difference in factors related to gender. Such committees will be key contributors to ensuring thorough supervision and oversight to make sure AI and ML can be used in a respectful, fair, and beneficial way for every patient, regardless of their gender [30].

Education and Training: It is critical to implement education and training programs that will deliver knowledge and comprehension of issues associated with the ethical application of artificial intelligence in the healthcare field to help understand complex issues relative to gender. As such, the main idea behind the program is to provide an understanding of how and when AI tools should be used allowing the healthcare professionals to familiarize with tools' capabilities and shortcomings. Moreover, it appears justified to include the aspects of AI impact on gender-focused care, which will empower the healthcare providers to better comprehend the nature of the disease when working with patients of various genders. In such a way, responsible utilization of AI does not perpetuate existing biases and misconceptions but creates the atmosphere of inclusiveness and awareness. In other words, it allows for more sensible gender-focused care. Finally, there is a need for a balance between the potential benefits of at and a multidimensional and multi-faceted vision of ethics. On the one hand, artificial intelligence and machine learning offer extensive opportunities relative to gender-focused care. On the other hand, human society should bear in mind the responsibility which necessarily arises from the count question of implementing a I in various ways to preserve a balance.

14.5 CONCLUSION

The chapter ends with a final reflection on the future directions for AI and ML in gender-based healthcare. It is clear that there is a need to look ahead, and the reflection prompts an even more important thought, that is, the need to ensure that the ethical frameworks and guidelines are developed with even greater sophistication. The perspective presented herein should not be understood solely as a theoretical exercise; actually, it is a call for a multi-disciplinary effort that marries technological innovation with a profound ethical engagement. AI and ML have an unprecedented potential to reengineer gender-based healthcare, improving care and outcomes for patients. However, they come with some equally unprecedented ethical challenges, which need to be deliberated and managed most meticulously. The world of digital healthcare is upon us, so professionals in healthcare and technology should team up with ethicists and policymakers. It is the joint endeavor of the four parties that will ensure the AI and ML technologies are implemented properly, in line with ethical principles. The matters such as privacy course, biases and fairness, and accountability need to be considered together. The most profound thought one should develop is the awareness of the fact that each of the stakeholders presents a unique perspective, which is needed to ensure that the future of healthcare is not simply technologically advanced but also ethically responsible and inclusive. The outlines of the ethical frameworks should be growing with the advancement of AI and ML technologies. Ultimately, to make sure gender sensitivity lies at the core of technological development, the AI systems should not only be unbiased but also responsive to different gender sensitivities. The ethical guidelines should promote inclusivity and fairness, making sure that the tools are developed and used with the awareness of the unique healthcare needs presented by different identities.

It further introduces the importance of transparency and explainability when it comes to AI and ML. With everything on the line in the healthcare industry such as people's lives, the decisions made by AI must be simple enough to be explained effectively. When everything is made known to the patients regarding the system run by AI-generated suggestions and decisions, they get confidence. Also, healthcare workers become more involved and become super responsible because they have the power to decide backed up by AI's opinions. Policy development is vital in the future of AI and ML in gender-related healthcare. Policies should seek to create a conducive environment that promotes and supports ethical AI research and application without violating the rights of the patients. Regulation should be moderate to foster innovation and to avoid bringing in unethical laws that may restrict the rise of technology.

Besides, Ethical training and education are crucial. Everyone has to be taught how AI is integrated into a system. The skills are also included for healthcare providers as vital members who will need to run systems in hospitals. It is a combination of AI education and ethics to see how safe AI will be in the future when being used to make decisions. In sum, the chapter provides a comprehensive and insightful perspective on the use of AI and ML in gender related to healthcare. It celebrates the potential of these tools to revolutionize healthcare while cautioning against the pursuit of advanced technology without a proper necessary ethical framework. By outlining a vision of the future of healthcare that is fairer and more productive, the chapter establishes an agenda for a future where technology is prioritized and guided by ethical principles. It is a matter of promoting AI in decision making that helps humanity while at the same time promoting caring persons through ethically making them be respected, caring individuals.

REFERENCES

[1] Kaur, G., Gupta, M., & Kumar, R. (2021). IoT based smart healthcare monitoring system: A systematic review. Annals of the Romanian Society for Cell Biology, 3721–3728.

[2] Gupta, M., Ahmed, S., Kumar, R., & Altrjman, C. (Eds.). (2023). *Computational Intelligence in Healthcare: Applications, Challenges, and Management*. CRC Press.

[3] Chandra, S., Saxena, A. (2021). Artificial intelligence and machine learning in healthcare. *Artificial Intelligence and Machine Learning in Healthcare*. Springer.

[4] Dahiphale, D., Shinde, P., Patil, K., & Dahiphale, V. (n.d.). Smart farming: Crop recommendation using machine learning with challenges and future ideas. https://doi.org/10.36227/TECHRXIV.23504496.V1

[5] Fishbain, B. (2021, September 26). Machine learning-based behavioral diagnostic tools for depression: Advances, challenges, and future directions. Journal of Personalized Medicine., 11(10), 957. https://doi.org/10.3390/jpm11100957.

[6] Gaur, L., Santosh, KC (2021). *Artificial Intelligence and Machine Learning in Public Healthcare.* Springer Briefs in Applied Sciences and Technology.

[7] Geigel, A. (n.d.). Machine learning AI systems and the virtue of inventiveness. AI and Ethics, 3(2), 637–645. https://doi.org/10.1007/S43681-022-00197-X

[8] Giovanola, B., & Tiribelli, S. (n.d.). Correction: Beyond bias and discrimination: Redefining the AI ethics principle of fairness in healthcare machine-learning algorithms. AI & Society. https://doi.org/10.1007/S00146-023-01722-0

[9] Islam, M. (2021). Smart healthcare in the age of AI: Recent advances, challenges, and future prospects. IEEE Access.

[10] Khare, A., Jeon, M., Sethi, I. K., & Xu, B. (2017). Machine learning theory and applications for healthcare. Journal of Healthcare Engineering, 2017, 1–2. https://doi.org/10.1155/2017/5263570

[11] Kearsley, L. (2006). Hybrid AI and machine learning systems. Computer, 2006–2007, 2–4.

[12] Kocheturov, A., Pardalos, P. M., & Karakitsiou, A. (2018). Massive datasets and machine learning for computational biomedicine: Trends and challenges. Annals of Operations Research, 276(1–2), 5–34. https://doi.org/10.1007/S10479-018-2891-2

[13] Lekkala, L. (2023). Artificial intelligence and machine learning based intervention in medical infrastructure: A review and future trends. UCjournals. https://doi.org/10.52402/Nursing207

[14] Lechterman, T. (2021). Combating disinformation with AI: Epistemic and ethical challenges. 2021 IEEE International Symposium on Technology and Society (ISTAS).

[15] Lau, N., Hildebrandt, M., & Jeon, M. (2020). Ergonomics in AI: Designing and interacting with machine learning and AI. Ergonomics in Design: The Quarterly of Human Factors Applications, 28(3), 3. https://doi.org/10.1177/1064804620915238

[16] Morley, J., Machado, C., Burr, C., Burr, C., Cowls, J., Taddeo, M., & Floridi, L. (2019). The debate on the ethics of AI in health care: A reconstruction and critical review. Social Science Research Network. https://doi.org/10.2139/SSRN.3486518

[17] Mujahid, O., Contreras, I., & Vehi, J. (2021). Machine learning techniques for hypoglycemia prediction: Trends and challenges. Sensors, 21(2), 546. https://doi.org/10.3390/S21020546

[18] M., H., Arnold. (2021). Teasing out artificial intelligence in medicine: An ethical critique of artificial intelligence and machine learning in medicine. Journal of Bioethical Inquiry. https://doi.org/10.1007/S11673-020-10080-1

[19] Nithya, N., Dixit, A. K., Billewar, S., Bobade, S. D., & Mohammed Alsekait, D. (2022). Machine learning based healthcare monitoring using 6G technology. Journal of Pharmaceutical Negative Results, 553–562. https://doi.org/10.47750/PNR.2022.13.S09.061

[20] Nikam, Swati Subhash. (2019). Prevasive healthcare and machine learning algorithms. International Journal of Computer Sciences and Engineering, 7(7), 32–34.

[21] Nuño, J. C. (2021). Machine learning for risk assessment in gender-based crime. ArXiv.

[22] Okon-singer, H. (2021). Machine learning-based behavioral diagnostic tools for depression: Advances, challenges, and future directions. Journal of Personalized Medicine, 7(7), 32–34.

[23] Pasricha, S. (2023). AI ethics in smart healthcare. IEEE Consumer Electronics Magazine. https://doi.org/10.1109/MCE.2022.3220001

[24] Peña, A., Serna, I., Morales, A., Fierrez, J., Ortega, A., Herrarte, A., . . . Ortega-Garcia, J. (n.d.). Human-centric multimodal machine learning: Recent advances and testbed on AI-based recruitment. SN Computer Science, 4(5). https://doi.org/10.1007/S42979-023-01733-0

[25] Publication, I. J. R. A. S. E. T. (2022). A machine learning based healthcare diagnostic model. International Journal for Research in Applied Science & Engineering Technology (IJRASET), 10(V), 3195–3199. https://doi.org/10.22214/ijraset.2022.42985

[26] Sen, R. (2021). Machine learning in finance-emerging trends and challenges. ArXiv (Cornell University).

[27] Shehata, S. (2021). Smart healthcare in the age of AI: Recent advances, challenges, and future prospects. IEEE Access.

[28] Sunagar, P., Hanumantharaju, R., Kumar, D. P., Sowmya, B. J., Seema, S., & Kanavalli, A. (2021). Smart healthcare: Using IoT and machine learning-based analytics. Studies in Big Data, 307–329. https://doi.org/10.1007/978-981-16-0415-7_15

[29] Yadav, K., & Gaurav, A. (2023). Application and challenges of machine learning in healthcare. International Journal for Research in Applied Science and Engineering Technology, 11, 458–466. https://doi.org/10.22214/ijraset.2023.55678

[30] Parashar, G. (2021). Systematic mapping study of AI/machine learning in healthcare and future directions. SN Computer Science, 2(6), 461. https://doi.org/10.1007/s42979-021-00848-6.

[31] Khankhoje, R. (2023). AI-Based test automation for intelligent chatbot systems. International Journal of Science and Research (IJSR). https://doi.org/10.21275/SR231216065308

Index

Printed in the United States
by Baker & Taylor Publisher Services

Printed in the United States
by Baker & Taylor Publisher Services